Springer Series in Biophysics

Editor: P. M. Bayley, London

7

Springer Series in Biophysics

W. R. Taylor (Ed.)

Patterns in Protein Sequence and Structure

With 88 Figures

Springer-Verlag

Berlin Heidelberg New York
London Paris Tokyo
Hong Kong Barcelona
Budapest

Dr. WILLIAM R. TAYLOR
Laboratory of Mathematical Biology
National Institute for Medical Research
The Ridgeway, Mill Hill
London NW7 1AA
United Kingdom

ISBN 3-540-54043-1 Springer-Verlag Berlin Heidelberg New York
ISBN 0-387-54043-1 Springer-Verlag New York Berlin Heidelberg

Library of Congress Cataloging-in-Publication Data. Patterns in protein sequence and structure / W. R. Taylor (ed.). p. cm. – (Springer series in biophysics; v. 7) "The contents of this
volume derive loosely from an EMBO workshop held at EMBL (Heidelberg) towards the end
of 1989" – Pref. Includes bibliographical references and index. ISBN 3-540-54043-1 (Berlin).
– ISBN 0-387-54043-1 (New York) 1. Amino acid sequence. 2. Amino acid sequence – Data
processing. 3. Proteins – Conformation. 4. Proteins – Conformation – Data processing. I.
Taylor, W. R. (Willie R.) II. Series. QP551.P35 1992 547.7′5 – dc20 91-42341 CIP

© Springer-Verlag Berlin Heidelberg 1992
Printed in the United States of America

The use of general descriptive names, registered names, trademarks, etc. in this publication
does not imply, even in the absence of a specific statement, that such names are exempt from
the relevant protective laws and regulations and therefore free for general use.

Typesetting: Camera ready by author
31/3145-5 4 3 2 1 0 – Printed on acid-free paper

Preface

The contents of this volume derive loosely from an EMBO workshop held at EMBL (Heidelberg) towards the end of 1989. The topic of *Patterns in Protein Sequence and Structure* attracted a wide range of participants, from biochemists to computer scientists, and that diversity has, to some extent, remained in the contributions to this volume.

The problems of interpreting biological sequence data are to an increasing extent forcing molecular biologists to learn the language of computers, including at times, even the abstruse language of the computer scientists themselves. While, on their side, the computer scientists have discovered a veritable honey-pot of real data on which to test their algorithms. This enforced meeting of two otherwise alien fields has resulted in some difficulties in communication and it was an aim of the EMBO workshop to help resolve these. By the end, most biologists at the meeting had, at least, heard the terms *Dynamic Programming* and *Regular Expression* while for their part the computer programmers began to realise that protein sequences might be more than simple Markov chains in a 20-letter alphabet.

Thanks to the modern facilities at EMBL, the three day meeting was video-taped and from this a transcript was taken and offered to the speakers as the basis for a contribution to this volume. For some of the more coherent speakers, the transcripts have been used almost verbatim, while others (who shall remain nameless) preferred to start afresh! Either way, most of the contributions still retain a directness that results from their oral presentation and it is hoped that this will aid in their easy assimilation by both biologists and computer scientists alike.

Of all who have been involved in the origins and production of this volume greatest thanks must go to the EMBO for funding the original workshop and assisting with some of the production costs. We are also greatly indebted to the EMBL for firstly, the use of their facilities and secondly, a copy of the video recording from which the transcripts were taken. The inception of this volume (as well as the workshop) was due to Peter Bayley (series editor), without whose encouragement little would have resulted. On the technical side, the rather tedious job of transcription was carried out with great speed and accuracy by Daphne Field while Christine Orengo has borne much of the burden of organising the contributions, including, along with Nigel Brown, their conversion into a more pleasing format using the (free) computer typesetting package LaTeX. Thanks go to all the above and also to the contributing authors for, either sooner or (more typically) later, providing a contribution.

The chapters follow, roughly ordered from pure sequence analysis to structure analysis, including, towards the end, even some experimental approaches. This progression is echoed by the gradual distortion of Marilyn Monroe's face into a protein motif — a piece of sacrilege (perpetrated by the author) which formed the poster advertising the original meeting. The poster was, of course, brightly coloured and those readers who have not exhausted their *Day-Glo* pens hi-lighting sequence motifs might like to reproduce the original effect by copying the colouring scheme of Andy Warhol's *Ten Marilyns*.

Willie Taylor
Mill Hill, London
1990

Contents

An Expert System for Secondary Structure Prediction

Patterns in Secondary Structure Packing — a Database for Prediction

Secondary and Supersecondary Motifs in Protein Structures

A Review of Methods for Protein Structure Comparison

Patterns of Sequence and 3-D Structure Variation in Families of Homologous Proteins: Lessons for Tertiary Templates and Comparative Modelling

T. L. Blundell **189**

Modelling From Remote Sequence Similarity — Enveloped Virus Capsid Structure Modelled on the Non-Enveloped Capsid

S. D. Fuller and T. Dokland **207**

Structural Motifs of the Extracellular Matrix Proteins Laminin and Tenascin

A Sequence Motif in the Transmembrane Region of Tyrosine Kinase Growth Factor Receptors

Introduction — Patterns, Predictions and Problems

W. R. Taylor

Laboratory of Mathematical Biology
National Institute for Medical Research
The Ridgeway, Mill Hill
London NW7 1AA
U.K.

w_taylor@uk.ac.mrc.nimr

The ultimate rationale behind all purposeful structures and behaviour of living beings is embodied in the sequence of residues of nascent polypeptide chains — the precursors of the folded proteins which in biology play the role of Maxwell's demons. In a very real sense it is at this level of organisation that the secret of life (if there is one) is to be found. If we could not only determine these sequences but also pronounce the law by which they fold, then the secret of life would be found — the ultimate rationale discovered!

Jaques Monod (1970)
from *Chance and Necessity* loosely translated from the French (and Latin).

The Ultimate Rationale

The revolution in molecular biology in the seventies began an explosion in the elucidation of biological sequence data that has revealed the sequences of many familiar proteins and many more new proteins, often of unknown function. One of the pressing problems in molecular biology is how to interpret these data to allow informed progress in the study of the proteins whose sequences have been determined. If a new sequence has a clear similarity to a protein of known function (and perhaps even structure) then much can be learnt very rapidly by simply recognising the homology. However, all too often a search across the sequence databanks returns no significant match, or perhaps only a match to an equally un-characterised protein. Faced with this situation, two lines can be pursued: one is to look for fragmentary similarities with other proteins rather than search for a similarity over the whole of the new sequence and the other is to attempt to predict the structure of the new protein. Both approaches rely on identifying characteristic sequence patterns and where possible, relating these to known structures.

Protein Structure Prediction

At first sight, the complexity of a three dimensional protein structure may give the impression of a randomly connected mass of atoms. However, a little perseverance and simplification will soon reveal some underlying organisation. Neglecting the side-chains of the polymer and concentrating only on the path of the back-bone, uncovers the overall fold of the chain, loosely referred to as its topology. The three hundred odd proteins of known three dimensional structure exhibit a wide variety of topologies but, surprisingly, there is little apparent connection between the protein fold and the nature of the amino acid sequence which overlays it and, to a large extent (if not wholly) determines it.

This enigmatic relationship between the protein sequence and its tertiary fold has been an endless topic of speculation and research ever since the first protein structures were determined (see opening quote from Monod above) and, allowing some limited successes, the basic problem is as great a puzzle today. The frustrating aspect of the problem is that, fundamentally, there is no mystery since the system is completely determined: we know the exact chemical structure of the protein, the nature of the solvent and have very successful theories that, for smaller systems, can predict chemical interactions with great accuracy. So, in effect, the 'folding laws' sought by Monod are known. Why then do they not work for proteins? The conventional wisdom answers that in principle they will but, as yet, we do not have the computer resources to run an accurate simulation for long enough. The difficulty arises from modelling the hydrophobic effect which is possibly the most important component in determining the overall tertiary fold. This is an entropic effect requiring sufficient bulk solvent around the protein chain to be simulated for perhaps tens of seconds (of the order of real folding times). Given that present simulations are measured in tens of pico-seconds, there is a gap of 12 orders of magnitude, giving little hope for answers from this approach in the near future.

Faced with frustration on the deductive front, many workers in the field turned to a more inductive approach. This attempts to circumvent the inaccessible kinetics of folding and considers only the final form. If correlations can be found between sequence and structure then the route that led to this correlation is irrelevant to its future predictive use. This aim, which is sometimes referred to as the empirical approach to protein structure prediction, was one of the original motivations behind the systematic analysis and correlation of sequence and structure. Unfortunately, the only obvious correlation between sequence and structure is that, for globular proteins at least, the more hydrophobic amino acids are predominantly found buried in the core while the more polar residues interface with the solvent. However, with a little work, correlations are also apparent between the sequence and the location of secondary structures and one of the early efforts in this direction was to correlate sequence patterns, principally of hydrophobicity, with secondary structure (Lim, 1974a,b). Recently further advances have been made on this front through the use of advanced computation techniques to resolve ambiguous predictions (Presnell *et al.*, in this Vol.) and also to eliminate some of the uncertainty through the use of multiple sequences (see Gibson, in this Vol.). Progress beyond secondary structure prediction towards the tertiary fold would be helped if further correlations between sequence and structure could be identified and given a large collection of motifs of known structure there is hope for some generality in this approach.

Motifs and Modelling

In more recent years the general prediction problem has been augmented by some less ambitious but more practical goals which have arisen through the large volume of sequence data that is now available and being produced at an ever increasing rate. These problems arise on the flanks of the *ab initio* structure prediction problem representing the extremes where either nothing is known about a new sequence and where the new sequence belongs to a well characterised family.

Given a new protein sequence about which nothing else is known, while it might be nice to know its 3D structure, it is perhaps more important, firstly, to assign a function to the protein. Since most protein sequences are determined by translation from a nucleic acid sequence this problem is now common and with the effort to sequence the human genome, is likely to grow. The simplest approach to assigning function is to compare the new sequence to all other sequences and hope for a similarity to emerge with a protein about which something is known. The required techniques are a sensitive method to match pairs of sequences, however, sensitivity costs time and with the growing databanks this can be lengthy. A more efficient alternative is to collect fragments, or sequence motifs, that characterise the known protein families and reverse the problem by scanning the motif collection against the new sequence (Taylor, 1986). This is a simple procedure for finding obvious relationships but for those that are more remote, difficulties can arise, requiring more sensitive matching tools such as the Template method of Taylor (1986, 1991) or the methods of Sibbald and Argos (in this Vol.) and Barton (in this Vol.).

At the other extreme, the new protein sequence may belong to a well characterised family and if there is a known structure for a member of this family then the structure prediction problem becomes almost trivially easy for the new sequence. Despite knowing the correct fold, there are still many problems to be solved in constructing a good model, and also the need to push the technique to the limit of recognisable similarity. As with sequence motifs, uncertainties can be reduced if there is more than one member of the family available since then it becomes clear where conserved features are located, which allows account to be taken in the modelled structure of why they might be conserved (see Blundell, in this Vol.).

Computational Techniques

The problems of assigning function, predicting structure and modelling by homology are all interrelated. For example: a sequence motif may have a known structure, such as the helix-turn-helix DNA binding motif (see Anderson, in this Vol.) which when identified, not only illuminates the function of the protein but also gives some structural information. If a chain fold can then be predicted, the technique of turning the backbone sketch into a plausible detailed model corresponds to the problem of modelling a sequence on the known structure of a homologue. Both of these three-dimensional modelling exercises require equivalent computational techniques based on interactive computer graphics in conjunction with a variety of energy optimisation methods. This overlap in computational technique also occurs on the sequence analysis side with the Dynamic Programming method of sequence alignment (Needleman and Wunsch, 1970) and more specialised pattern matching methods being used not only to find sequences that contain particular patterns but also produce a multiple alignment of the family and even for secondary structure prediction.

With a few exceptions, the techniques mentioned above can be referred to as standard in that the field of protein analysis is not principally concerned with their fundamental development and merely adopts them as tools. Some exceptions might include the Dynamic Programming algorithm of Needleman and Wunsch (which was discovered independently but had been used in economics since the 1950s) or its recent extension into three-dimensional analysis (see Orengo, in this Vol.). Despite not being at the cutting edge of algorithm development, the problems encountered in the analysis of sequence and structural data are not trivial. These data constitute a qualitatively complex mix that does not easily conform to a standard, say, statistical analysis. Thus, although the common tools are often standard, ingenuity is needed in their application.

Current Problems in Pattern Matching

As an example of the type of difficulties which can arise, consider the problem of defining a sequence motif: a problem that is central to the topic of this book. The common approach is to adopt some feature which has been identified by 'eye' as being characteristic and to match it against other sequences. This, I will refer to as the *anecdotal* approach. It can be argued that the anecdotal approach presents no problems providing a rigorous test is available to assess the results. For example, Pearl and Taylor (1987) developed a pattern to relate the retroviral proteases to those of the pepsin family and used the criterion that the pattern must match both families and no others in the sequence databank. Thus whatever the origin or form of the pattern, it embodied something common and unique to both families, so demonstrating their relatedness. This seems reasonable but is it reliable? It may, indeed, be possible to find something common and unique to any two families. To derive rigorous constraints on this rule-of-thumb for all sizes of families against all sizes of databank would be a daunting mathematical task, if not impossible, given the non-random nature of protein sequences.

Against the anecdotal approach is the drive to systemise the origin of patterns. This does not avoid the above problem of significance but can be justified both by the scientific method and by the time that will be saved as the databanks expand. Automatically manipulating patterns opens the way to refining them to be optimally discriminating for some feature or family. Firstly, it is necessary to adopt some definition of specificity based, say, on a function of true and false hits. There is a large field of theory on this topic (discriminant analysis) but to apply it requires that the distributions of scores are known, which for the current problem is generally not possible. A rough score of sensitivity might be defined as: wT/F, where T is the fraction of correct hits, F the fraction of wrong hits ($+1$) and w a weight to adjust the relative importance of the two, however, even with this definition settled there are still further problems.

To refine a pattern it is necessary to have a functionally homogeneous group of proteins that can be distinguished from the rest of the databank. While this may not appear to be a problem where a single protein family is concerned, when more general properties are considered then the multi-faceted nature of proteins can lead to complexities. Consider for example the well known glycine-rich nucleotide binding motif (see Hodgman, in this Vol.). This pattern was originally identified for di-nucleotide binding proteins but a similar pattern is also found in mono-nucleotide binding proteins (which includes many families of kinases and other GTP binding proteins). A pattern designed to be specific to one subgroup also identifies proteins from the other, while a common pattern identifies further potential members such as co-enzyme A binders (which has a nucleotide component) and even DNA associated proteins. The solution here might be to devise a hierarchy of patterns based on generality, as was attempted for the immunoglobulin domains (Taylor, 1986, 1991). This would entail, at the base level, a large number of quite specific patterns merging towards the most general pattern at the apex. Together these patterns would then be weighted so that if no specific pattern was located, the next most general would be tested.

The further, and perhaps more fundamental, problem then arises that with most protein sequences now being known only from the translation of a nucleic acid sequence, they are usually functionally classified only by patterns in their sequence. Thus the criteria of truth (about function) and the patterns that we hope to refine become interdependent. In practical terms this implies that the pattern matching tool must include a classification facility with which to redefine its own groupings. The problem then becomes more open and in the limit is equivalent to feeding the sequence databank into a program and asking "What interesting things can you find in there?". The assumption that the program understands the concept 'interesting' presupposes some intelligence on its part and while some may think this is now going too far, others, in particular Smith and co-workers have followed this logical path and begun applying the techniques of Artificial Intelligence (AI) to the problem. (Smith and Smith, 1990).

A subsidiary problem to this approach is to decide how much help to provide. For example, if an AI returned from a sojourn in the sequence databank with the conclusion that flavodoxin is a globin, then many crystallographers would feel that some source code needs to be rewritten. Alternatively, the AI could be given access to the structure databank, and in the end it can be argued that it should have access to all the knowledge sources that we humans have, including structural data, sequence annotation and even the text of relevant papers (for the meantime the line might be drawn at practical bench work!). Then, if nothing else, it will at least verify the integrity of our data by compiling a massive cross-reference. Without the rich mental background of a human investigator it is unlikely that the concept 'interesting' will be easily encoded, however, diffuse data structures such as neural networks are often surprising in their ability to capture ill-defined criteria and perhaps some feed-back on where the results get published (if at all) may be all that is required.

On the structural side, the problems of motif definitions and significance are equally fraught and further complicated by their more limited instances. As an example, consider the problems of simply defining the protein topology. This is often based on definitions of hydrogen-bonds which, especially in β-sheets, locate and specify the conformation of the chain fold in a discrete manner which is easily encoded. Unfortunately, some β-strands might be defined only by a few hydrogen-bonds and a shift of a fraction of an Ångstrom (well within normal error) might make the difference between the identification of the region of chain as a β-strand or an 'unstructured' loop. Loss of such a weak strand might in turn lead to a radically different classification of the topology of the fold. The problem is not new as the following adaption illustrates:

For the want of a bond, a strand was missed,
For the want of a strand, a sheet was missed,
For the want of a sheet, a fold was missed,
For the want of a fold, a similarity was missed,
For the want of a similarity, a publication was missed,
All for the want of a hydrogen-bond.

(with apologies to anon.)

While some of the recent 'soft' methods of structure comparison might overcome these difficulties (see Orengo, in this Vol.) there still remains a problem of how protein folds should be classified. The definitive work on the subject was based largely on visual appraisals (Richardson, 1981) and although some attempts have been made to automate topology identification (e.g. Rawlings *et al.*, 1987), this area of expertise still lies in the control of humans. Similarly, the classification of motifs relies on careful studies which are largely unautomated (see, for example, Thornton *et al.*, in this Vol.). These studies, however, define a methodology and provide necessary examples that more electronic counterparts might well aim to emulate in the future.

8

Conclusions

The above discussion gives an idea of some of the more basic outstanding problems in the field of sequence and structure pattern matching and it is likely that tackling these can only become more difficult as the volume of data increases. While progress is clearly being made on the motif and modelling fronts, these have not yet met sufficiently in the middle ground to provide a general structure prediction method. Either the motifs are too small to give an idea of the overall fold or else correspond to the whole protein which can then be directly modelled. It is difficult to predict whether this situation will be resolved as more motifs are identified, however, there is hope that the number of fundamental motifs may be limited. This hope comes from the observation of equivalent motifs in otherwise unrelated proteins, such as the barrel structure found in Triosephosphate IsoMerase (TIM) and at least a dozen other distinct proteins, and also from the operational difficulties faced when evolving new proteins. Even from our own (human) experience, it has been found to be very difficult to design a protein from scratch but relatively simple to tinker with an existing structure. Similarly, proteins have evolved from each other in this way and, it might be supposed, ultimately derive from a limited number of basic precursor structures. It has been proposed that these fundamental units correspond to exons and that they may be limited in number to as few as several thousand (Dorit *et al.*, 1990).

We can hope that as both the sequence and structure databanks grow, more sequence motifs will be identified from new families and new links forged between previously unrelated families. As connections grow through the sequence data, so more motifs will be identified with both old and new structures. One of the hopes of protein data analysis, and pattern matching studies in particular, is that eventually such connections will span the entire sequence databank linking every sequence pattern home to a three dimensional structure. If this is achieved then although we may not have found Monod's folding laws, we might still hope to predict structure from sequence. The only question then remaining is whether this solution counts as discovering the Secret of Life?

References

Dorit, R. L., Schoenbach, L., and Gilbert, W. (1990). How big is the universe of exons? *Science*, 250:1377–1382.

Lim, V. I. (1974a). Algorithms for the prediction of α-helical and β-structural regions in globular proteins. *J. Mol. Biol.*, 88:873–894.

Lim, V. I. (1974b). Structural principles of the globular organisation of protein chains. A stereochemical theory of globular protein secondary structure. *J. Mol. Biol.*, 88:857–872.

Needleman, S. B. and Wunsch, C. D. (1970). A general method applicable to the search for similarities in the amino-acid sequence of two proteins. *J. Mol. Biol.*, 48:444–453.

Pearl, L. H. and Taylor, W. R. (1987). A structural model for the retroviral proteases. *Nature*, 328:351–354.

Rawlings, C. J., Taylor, W. R., Nyakairu, J., Fox, J., and Sternberg, M. J. E. (1985). Reasoning about protein topology using the logic programming language PROLOG. *J. Mol. Graphics*, 3:151–157.

Richardson, J. S. (1981). The anatomy and taxonomy of protein structure. *Adv. Prot. Chem.*, 34:167–336.

Smith, R. F. and Smith, T. F. (1990). Automatic generation of primary sequence patterns from sets of related protein sequences. *Proc. Natnl. Acad. Sci. USA*, 87:118–122.

Taylor, W. R. (1986). Identification of protein sequence homology by consensus sequence alignment. *J. Mol. Biol.*, 188:233–258.

Taylor, W. R. (1991). A template based method of pattern matching in protein sequences. *Prog. Biophys. Mol. Biol.*, 54:159–252. In press.

A Brief Review of Protein Sequence Pattern Matching

David T. Jones

Biomolecular Structure and Modelling Unit
Biochemistry and Molecular Biology Dept.
University College
Gower Street
London WC1E 6BT
U.K.

jones@uk.ac.ucl.bioc.bsm

Introduction

Of the many current and abandoned methods for the prediction of a protein's tertiary structure from its amino acid sequence, the diverse yet related set of pattern matching techniques offer, perhaps, the best potential for solving this fundamental problem in molecular biology. The need for reliable structure prediction techniques has never been greater, and with the future prospect of sequencing the entire human genome, the necessity of a workable solution can only become significantly more acute.

To understand the reason for studying pattern matching in the field of protein structure prediction it is of course necessary to consider the more obvious approaches that have been tried with varying degrees of success. The general topic of protein structure prediction is far too big to discuss here in anything but cursory detail, yet this survey would be incomplete without some consideration of the traditional attempts at solving the 'folding problem' as it has come to be known.

Evidence for a direct relationship between the primary structure of a protein and higher levels was first presented by Anfinsen *et al.* (1961). It is generally thought that the amino acid sequence contains all the information required to permit a polypeptide to 'self-assemble' itself into a biologically active three-dimensional structure. Though this folding of the protein is context sensitive, in that its correct execution is highly dependent on the ambient conditions and the possible presence of cofactors or even 'molecular chaperones', the mechanisms that decode the structural information are in principle very simple. The structure of a folded protein is maintained through a complex interplay between a handful of physico-chemical forces: electrostatic effects, hydrogen bonding, hydrophobic effects, solvent entropic effects, salt bridging and so on. Most of these forces are well defined and have a solid backbone of theory behind them. In theory, the effects of hydrogen bonding can be modelled by simply solving the relevant quantum mechanical equations, solvent effects modelled by molecular dynamics, the list could continue. Predicting the structure and function of a protein sequence is therefore conceptually simple. A set of differential equations could be constructed where each physical force is simulated and the folding process itself simulated as a result. Given a sufficiently powerful computer an acceptable solution could be found. However, the number of these simultaneous equations that must be solved to predict the final protein conformation is astronomical. Though the number of atoms in the polypeptide chain itself is small, it is also necessary to consider the multiple interactions between the surrounding solvent molecules and the chain. By using suitable simplifications some progress has been made along these lines, however even where theoretical solutions have been within reach, it has proven difficult to find the correct energy functions and even more difficult to find functions that allow a *convergent* solution to be found [see for example Levitt (1976) and Robson and Osguthorpe (1979)]. Suffice it to say that though a direct theoretical approach to solving the folding problem could be envisaged, the problem may well turn out to be intractable via this route alone.

Given that we cannot predict protein structure from first principles, other 'heuristic' methods must be found. An obvious approach to solving vastly complex systems of equations is to merely observe the macroscopic properties exhibited by a wide range of different final solutions. In this case, a sensible approach is to analyze the final folded states of different proteins statistically. Amongst the attempts at analyzing the relationship between protein sequence and structure statistically were those of Chou and Fasman (1974) and Garnier *et al.* (1978). These attempts were strictly aimed at predicting the secondary structure of proteins. The basic idea behind these techniques is to assign a structural 'propensity' to either individual residues (e.g. Chou and Fasman) or short sequence segments (Garnier *et al.*). Though in a sense a degree of pattern matching is being carried out in these methods, they are not strictly pattern matching approaches. For a review of these essentially statistical approaches see Thornton and Taylor (1989).

Pattern Matching Methods in the Prediction of Protein Structure and Function

Taylor (1988a) explains that the applicability of different pattern matching methods to given situations depends on the degree of homology found between the sequence under scrutiny and the database of known structures. For cases where the percentage identity (percentage of conserved residues between sequences) is 50% or above the case is clear-cut. The function of a protein will in all probability have been fixed by the location of such a close homologue. The structure will also follow closely, using the available techniques for modelling homologous proteins [e.g. Blundell *et al.* (1987)]. Highly conserved residues can be spotted easily by the construction of suitable conservation plots, and using this information to 'pin' the unknown structure to the known structural 'template' a reasonable model can be produced. This approach is currently the most successful structural prediction method available to date.

Where a significant degree of homology ($>$ 50% identity) exists between two sequences, pattern matching merely requires their optimal alignment. Two classes of global sequence comparison techniques exist: one based on the concept of *dynamic programming*, the other based on the identification of common subsequences. Dynamic programming provides the most general and rigorous solution to this particular problem. The application of dynamic programming methods to the alignment of biological sequences was first described by Needleman and Wunsch (1970). Alignment begins by the construction of a similarity matrix thus:

	A	C	D	E	F	G	H	L
A	<u>1</u>	0	0	0	0	0	0	0
D	0	0	<u>1</u>	0	0	0	0	0
D	0	0	1	0	0	0	0	0
E	0	0	0	<u>1</u>	0	0	0	0
F	0	0	0	0	<u>1</u>	0	0	0
G	0	0	0	0	0	<u>1</u>	0	0
P	0	0	0	0	0	0	0	0

In this simple case exact residue matches are given a score of 1, and any other match a score of zero. More 'lenient' scoring schemes are commonly used, particularly that of Dayhoff (1978).

The Needleman and Wunsch method continues by the dynamic calculation of all possible paths through the matrix starting at the bottom right hand corner, finishing at top left. Values in the similarity matrix are replaced by the maximum score obtainable *from that point on*. At the end of the summing procedure a maximum value should be found along the top row or the leftmost column which indicates the starting point for the optimal alignment path (the optimal alignment path for the above matrix is shown by the underlined scores). A further refinement to the method is the addition of a gap penalty which prevents the insertion of a ridiculous number of gaps in order to maximize the overall score.

Dynamic programming, though providing a rigorous, optimal alignment, suffers from a severe lack of speed, even when implemented on a fast computer architecture. Though several attempts have been made to improve the performance of the standard Needleman and Wunsch methods [see Taylor (1988b)], faster approximate alignment techniques are more commonly used for 'front-line' pattern matching with large databases. These approximate techniques are all based on the identification of common subsequences or 'tuples'. The original technique based on tuple comparison was the crude but effective 'dot-matrix' technique [see Staden (1982) for a review]. In this technique, the two sequences are written along the x and y axes respectively of a matrix, and a dot placed at each location where the corresponding residue pairs match. The resulting matrix is inspected visually in order to detect diagonal runs of dots which are indicative of homologous stretches between the two sequences. Though the output from the dot-matrix technique is nicely visual, it is obviously not an automatic technique.

Automatic tuple comparison programs have been produced, notably by Wilbur and Lipman (1983) and Lipman and Pearson (1985). The FASTx programs of Lipman and Pearson [see Pearson (1990) for a review] have become a *de facto* standard in the front-line database searching field. In all these methods the principle is simple. Each sequence is split into its respective tuples (subsequences of lengths 2..n) and these subsequences stored in a single hashed lookup table. The structure of this table is such that tuples found to be in both sequences are stored in a linked list. An alignment is produced by locating *significant diagonals*, which are simply the diagonals of the dot-matrix that have an above-average number of common tuple matches. As already stated, the main advantage of these methods is their speed: performance being around 40–50 times greater than that of dynamic programming approaches. The major disadvantage with tuple methods is that they rely on finding a fairly high number of common tuples, which in the case of fairly dissimilar sequences ($< 50\%$ ID) will not often be the case. Related to this problem, is the difficulty in providing a decent 'similarity' scoring scheme where amino-acid similarity is scored, rather than amino-acid identity.

Simple pairwise alignment methods as described above are good at comparing sequences where the homology exceeds 50% or so. In some cases two sequences may in fact have an *overall* homology much less than 50% and yet have a common segment with extremely high homology. Terminal and loop regions may well be highly variable in sequence, or may even be deleted or extended across a single protein family. In these cases a global alignment across the full length of both sequences may fail to provide a significant result. Both the dynamic programming and tuple alignment techniques have been modified to allow the identification of regions of maximum *local* homology. In the case of dynamic programming, the single-stretch best local alignment method of Smith and Waterman (1981) has proven popular, however a more complicated method by Sellers (1974, 1984) allows multiple subsequences to be detected in one operation. For an example of tuple-based local alignment see Waterman (1986).

Consensus Methods

As the degree of homology between two sequences drops below 50% it becomes difficult to locate the *biologically* optimum pairwise alignment between them. Usually, however, more than two sequences are available. Given several examples from a single family of proteins it is possible to construct a consensus alignment between them. The principle of *consensus* methods is seen mirrored in many fields of science and mathematics. In an abstract sense, consensus alignment is simply an example of *statistical sampling*. The basic idea is that instead of a single amino-acid code at each alignment position, a histogram is constructed where the numbers of each of the 20 amino acids occurring at that position are tallied. This allows each sequence to 'vote' for the appropriate residue at any given alignment position. Usually the total alignment is iterated, using the consensus patterns built in the previous pass to direct the alignments of the following pass. The concept of consensus pattern matching will be discussed in more detail later.

Multiple consensus alignment techniques are useful for average homologies above around 30% at which point all alignments become statistically insignificant. The 30% identity cut-off very roughly marks the outermost boundary of the so-called 'Twilight Zone' [Doolittle *et al.* (1986)]. The Twilight Zone is simply defined as the region of homology where an optimal alignment between random sequences is found to be no worse or better than the alignment between the trial protein sequences. It is unfortunate that in many documented cases useful alignments lie well inside the Twilight Zone. The problem is simply down to the extreme variability of protein sequence even where the higher levels of structure are seen to be highly conservative. Two very different sequences can quite easily have extremely similar folded conformations. Of course we might reasonably expect certain key residues to be conserved across the evolutionary tree, but how can these key residues be located amid the extreme evolutionary 'noise'. The number of key residues may well be very small in comparison to the lengths of the sequences, but methods have been developed to enable such residues to be rapidly located. See Gribskov *et al.* (1990) and Taylor (1990) for a fuller description of consensus alignment techniques.

The methods employed to detect remote homology between sequences are the truest forms of pattern matching. The earliest example of a strictly pattern matching approach to sequence comparison was that of the helical wheels of Schiffer and Edmundson (1967, 1968). Schiffer and Edmundson studied the distribution of hydrophobic residues along the lengths of α-helical sequence segments by plotting each residue in a circular fashion corresponding to the pitch of the helix. α-helical regions generally exhibit clustering of hydrophobic residues along a single sector of the wheel. Though of some vintage, this technique remains a powerful means for identifying and comparing α-helical regions. The method of helical wheels was refined by Palau and Puigdomenech (1974) who analysed the zonal distribution of hydrophobic residues in helical regions. They found that in helical regions hydrophobic residues at sequence offsets n, n+1 and n+4 or n, n+3, n+4 acted to stabilise the helices.

Shortly after Palau and Puigdomenech's article two articles by Lim (1974a,b) were published. The first paper presented a general study of the stereochemistry of globular proteins, showing which residue types could be associated with regular (α or β) or irregular protein conformations. In the second article, observations made on proteins from the contemporary structural database (25 structures in total), as outlined in the first article, were distilled into 22 rules. These rules tended to be verbose and highly general. For example, Rule 3 for α-helix formation is as follows:

> *"Let the hydrophobic pair (1-2) or the hydrophobic triplet (1-2-5) formed from positions (i, i+1) and (i, i+1, i+4) respectively, and position i+1 contain phenylalanine. Such a hydrophobic pair (1-2) or a hydrophobic triplet (1-2-5)... will be α-helical if Phe situated in position i+1 forms within the limits of the obtained α-helix a hydrophobic-hydrophillic pair (1- 5) with GL which is situated in position i-3."*

Lim claimed a prediction accuracy of 80–85% when these rules were applied to the same structures on which the rules were based. Though the accuracy of Lim's method is certainly not as high as 80% when applied to proteins absent from the original data set, it does in fact score higher, on average, than the more popular statistical techniques. Not surprisingly rules of this type were not easy to convert into computer code, which contributed to their lack of popularity. Taylor (1988a) points out that modern logic based languages seem to be suitable vehicles for encoding Lim's rules, and that work along these lines has been started by C. Rawlings and P. Stockwell (personal communication).

All the above pattern matching methods are concerned with detecting simple patterns primarily designed to predict secondary structure. The trend in protein sequence pattern matching has moved towards the construction of more complex templates, capable of matching higher levels of structure than basic α-helices or β-sheet regions. Given that long-range interactions play an important part in the direction of protein folding, even at the purely secondary structural level, the necessity for these complex patterns is reasonably obvious. Nagano (1977) first described the use of a super-secondary structural motif, where the $\beta\alpha\beta$ unit was analysed. Following on from this work Nagano (1980) extended the algorithm to a generalised structure prediction system. This approach splits the sequence into pentapeptides to reduce the number of degrees of freedom in the folding simulation. The folding is further constrained by considering the packing in a 2D matrix (3×11 boxes) rather than the 3D atomic coordinate space. The nub of the method is simple in that each pentapeptide is labelled as being α or β depending on the standard secondary structure prediction probability for each. Likely $\beta\alpha\beta$ units are then located by considering pentapeptide patterns that neighbour strongly predicted α or β segments, with an appropriate distance filter that excludes β-α/α-β pairs that are too far apart on the grid. An important part of this method was a *combinatorial* analysis of all the possible permutations of predicted structural segments, this is analogous to scanning through all the possible three-dimensional packings given a mixture of well-defined and ill-defined structural units, except that in this case packing is performed on a 2D grid.

Richmond and Richards (1978) described a fairly involved technique for tertiary structure prediction. Though the method used a wide range of techniques outside the realm of pattern matching, a pattern matching approach was used at the core of the method to identify hydrophobic residues important for the packing of secondary structural elements. The number of packing permutations matching the given patterns tended to be enormous, though this number was quickly reduced by means of simple distance filters applied in particular to the ends of helices. Richmond and Richards only considered $\alpha\alpha$ proteins (e.g. myoglobin) where the hydrophobic patterns were well defined, and the distance filters easy to construct and apply. Cohen *et al.* (1980, 1982) extended this combined pattern-matching and packing technique to $\beta\beta$ and $\beta\alpha$ folding types. This extension demanded the construction of hydrophobicity patterns for β structure similar to those already well-recognised for α-helices. Success in this case depended on the provision of some external knowledge to the prediction problem. The fundamental limitation here is the lack of accuracy in secondary structure prediction. Of course, given knowledge of the protein's folding type, secondary structure prediction can be weighted towards reasonable accuracy, and indeed where this information is available, the method of Cohen *et al.* provides at least a fair guess at the overall protein topology.

The methods of Richmond and Richards and Cohen *et al.* essentially fail due to a lack of accuracy in secondary structure prediction. The above methods are combinatorial and as such any uncertainties in the elements playing a part in the prediction process are grossly magnified by the time a complete tertiary structure is produced. Reducing the vast number of possible packing permutations relies on the provision of suitable filters applied to the component secondary structural elements. If the secondary structural elements are ill-defined (or rather mis-predicted) then we should not be surprised if the final predicted state is far from reality.

Taylor and Thornton (1983, 1984) attempted to improve the accuracy of secondary structure prediction by constructing templates capable of detecting super-secondary structural elements, or more specifically (but not exclusively) the $\beta\alpha\beta$ unit. Using 62 examples of the $\beta\alpha\beta$ unit from the Brookhaven database an ideal secondary structure sequence template was constructed. This ideal $\beta\alpha\beta$ pattern was matched at each residue position of the test sequence matching the template profiles to the Garnier secondary structure prediction probability profiles [see Garnier *et al.* (1978)], and a score calculated. Different length variants of the ideal template were created by scaling the master template so as to accommodate the length variations observed in the available examples. Apart from the statistical sequence template, templates were also constructed for matching patterns of hydrophobicity (as in Lim's method) — one template scored highly for buried β regions, the other for α-helical regions. The strongest fitting template was selected, and other matching templates selected according to various rules (for example forbidding overlapping α and β regions). On a test set of 16 β/α proteins, a prediction score of 70% was achieved, bettering the raw GOR secondary structure prediction technique by some 7.5%.

As an extension to the work on $\beta\alpha\beta$ templates, Taylor (1986) went on to produce a generalised consensus template method. The first major improvement to the original template method of Taylor and Thornton was to move over to 2D templates that could match more than one physico-chemical criterion at each alignment position. The second major improvement was to contrive a means for generating the templates automatically, given a suitably well-defined sequence alignment. The method starts with a seed alignment, generally based on available structural information. A consensus pattern is created from this initial alignment such that each alignment position contains a count of each of the 20 amino acids. This ties back to the earlier discussion of multiple sequence alignments using consensus methods. One important advantage of consensus patterns is that they are insensitive to the odd misalignment. Consider the case where, for example, in 10 alignments a glycine residue is found at a particular alignment position and in the eleventh, due to a misalignment, a proline is aligned with the glycine consensus. Evidently, the proline match will be recorded, but in future alignments, the proline will be scored 10 times lower than the glycines — the 'glycine-like' tendency of that particular template position will be more-or-less conserved.

The initial consensus template, preferably solidly based on structural knowledge, is then used to collect further matching sequences from the sequence database. The new set of sequences is then aligned to the consensus, to produce a new expanded consensus template, ready for a further cycle of sequence collection, alignment, and template generation. When no new sequences are collected, the alignment cycle is exited and the final minimal set template constructed. The nature of this final template is totally different from the consensus templates used to direct the collection/alignment cycle. Rather than recording the number of each of the 20 amino acids observed at each position, the observed amino acid identities are used to pick a minimal covering class of amino acid from a Venn diagram. This Venn diagram comprises three major sets: Hydrophobic, Polar and Small (with subsets Aromatic, Aliphatic, Tiny, Charged and Positive). The smallest subset that contains all the residues observed at a particular alignment position is selected, and is recorded in the final template at the corresponding position. There are two main advantages to using minimal covering classes over consensus patterns. Firstly, minimal covering class patterns are *predictive* in that they are able to predict possible amino acid substitutions that may not have been observed in the limited data-set that was used in the template's construction. The second advantage is that bias in the data-set is eliminated. We now return to our previous case of 11 alignments where one proline is matched against 10 glycines and if this alignment proves to be valid and indeed a proline *is* a valid residue at that position then in the final template we *do not* wish to score it lower than the more common glycine residues. Indeed, it may turn out that the sequences that contributed the 10 glycines at that position were in fact highly homologous and that, were the sample of sequences less biased we might observe just as many cases where proline is present as glycine. Taylor's method concludes by aligning each of the contributing sequences against the final template using a set-based scoring scheme rather than a consensus-based scheme. If the whole process has been successful then we expect the set-based alignment to concur with the consensus alignment.

Pearl and Taylor (1987) used the above consensus template program suite to identify common features in the retroviral protease and aspartyl protease families. The method was able to detect the few conserved residues that formed the proposed active site even though the sequences were extremely non-homologous with respect to normal alignment methods.

To enhance their previous work on secondary structural packing, a pattern matching approach to predicting structure in β/α proteins was formulated by Cohen *et al.* (1983). Their algorithm concentrated on turn prediction, achieving a prediction score of 98%, though by using the reliable turn prediction a complete structure prediction was produced as a final result. The method is essentially based on Lim-like patterns, though by considering segments separated by predicted turns, further constraints could be applied to cope with ambiguous predictions. The first stage of the method (TURNGEN module) involves locating turns by their characteristically high polar residue content. These predicted turn regions delimit independent sequence segments that are to be assigned as α, β or 'null' (null structures can be irregular, or just isolated regular structures that do not interact with the central β-sheet). Using a large array of small patterns, each delimited region is analysed, and labelled as possibly α or β depending on the overall pattern matching score (ABGEN module). The next stage attempts to label some segments as being *definitely* α, β or null (ADEF, BDEF and NULLDEF modules) by using pattern combinations (for example if a segment has been previously labelled as a potential α-helical segment and the hydrophobic diamond pattern 'S' matches, then the segment is assigned as definitely α). Remaining uncertainty in the prediction is resolved by the application of high-level 'expert system' rules containing well-founded knowledge on the structure of β/α proteins (ADJUST module). Processing continues (DELIMIT module) by using specific N and C terminal patterns to try to accurately determine the boundaries of the secondary structural elements now known to be contained in the 'definite' segments (for example a 'stop signal' for α-helices might well be one or more prolines or three hydrophobic residues). The final prediction is achieved by means of a scoring and ranking procedure (SCORE module) and by another expert system analysis of all the remaining combinatorial possibilities of definite segments and possibles (COMBINATORICS module). This combinatorial step is somewhat similar to the method of Nagano (1980) as described previously. The final icing is provided by filtering out the remaining statistical anomalies (OUTLIER module).

The method of Cohen *et al.* described above is complicated, but is interesting in that it uses earlier pattern matching approaches, using a rigorous artificial intelligence approach to handle ambiguous predictions. Another very interesting aspect is the use of expert rules to direct the global prediction process. Such a combination of methods will no doubt play a vital part in successful future prediction schemes, drawing predictive power from both blind pattern matching methods, and the hard-earned knowledge of experts in the rules of protein structure. From current evidence it would appear that prediction techniques drawing from one approach or the other (but not both) are only achieving prediction scores of 60–70% or so.

Regular Expressions

Another type of protein sequence pattern that has recently become popular is that based on *regular expressions*. A regular expression is simply a linear sequence pattern that permits the use of wildcards (matching any residues), set closures (matching residues in a particular set), gaps (matching a number of residues or none at all). To fit in with the complete definition of a regular expression, it must also be possible to define nested sub-patterns. For example in the pattern XYZ(P(QR))XYZ, the possible matches are XYZPQRXYZ or the sub-pattern PQR, or the sub-sub-pattern QR.

Probably the earliest example of true regular expressions being used for matching complex biological sequence patterns was the QUEST system designed by Abarbanel *et al.* (1984); Fig. 1. QUEST is a rapid search tool that implements a concise pattern language, very closely modelled on the syntax of the Unix EGREP (Extended Global Regular Expression search and Print) program. QUEST was designed to be able to handle the kinds of patterns thought useful in sequence analysis and structure prediction, and allowed patterns to be defined in terms of named or explicitly defined sub-patterns. For example in turn-prediction a sub-pattern would simply be the set of single residues with a predilection for turn-formation ([PGQNSTED-RKH]) called for example 'tphilic'. A further sub-pattern would be [YPGQNSTEDRKH] (tphilic plus

QUEST REGULAR EXPRESSION SYNTAX

.	- Any character/residue
[PQR...]	- Any character/residue in list PQR...
[~PQR...]	- Any character/residue NOT in list PQR...
*	- Matches zero or more occurrences of preceding item
+	- Matches one or more occurrences of preceding item
?	- Matches one or zero occurrences of preceding item
(regexp)	- Groups together regular expression as a single item
OR	- Matches preceding item OR following item
AND	- Matches if preceding and following items match
&	- Matches if preceding THEN following item match
^	- Matches start of sequence
$	- Matches end of sequence
{m}	- Matches previous item m times
{m,n}	- Matches previous item m..n times

FIGURE 1. QUEST regular expression syntax.

tyrosine) called 'yturnphilic'. A QUEST meta-pattern for a turn could then be defined as "tphilic3 yturnphilic", or in English : three turn-formers, plus one turn-former or tyrosine.

A derivative of QUEST (which was coded in the commercial IBM pattern language MAINSAIL) called PLANS was used by Cohen *et al.* for further work on turn prediction. The derivative pattern system is known as PLANS (Pattern Language for Amino and Nucleic acid Sequences), and was coded in Lisp. With the convenience of a well-tailored pattern matching language, Cohen *et al.* analysed turn patterns in all structural classes of proteins : $\alpha\alpha$, $\beta\beta$, $\beta\alpha$ and $\beta + \alpha$. By virtue of the QUEST/PLANS meta-pattern capabilities, the pattern library was greatly extended and a high prediction accuracy was achieved (95% over all examples, 90% on homologous proteins extracted from the sequence database). On the debit side, this accuracy was achieved by the inclusion of many parameters based on global knowledge extracted from all available protein structures, so until a sufficiency of new structures becomes available on which to test the algorithm in an unbiased fashion, the true accuracy remains uncertain. Nevertheless, the current indications are that this method may very well provide an impressively powerful means for turn prediction (especially if the folding type of the protein can be independently determined).

Specific Structural Patterns

All the above methods are essentially generally applicable automatic methods for protein pattern analysis. Although some of the methods are more generally applicable than others, none of them attempt to apply rigid pattern matching rules based on comprehensive structural knowledge. Of course, pattern matching of this sort requires one of two things: either the structural unit is extremely simple, or the structural unit is fairly high-level (but then limited to a specific family of proteins).

The earliest analysis of a specific protein fold was that of the dehydrogenase nucleotide binding fold by Wooton (1974). The subject of nucleotide binding motifs has been repeatedly studied time after time. Well known work in this area includes the ATP binding patterns of Walker *et al.* (1982), and the $\beta\alpha\beta$ dinucleotide binding motif of Wierenga and Hol (1983). Other 'macro-patterns' include calcium-binding patterns [for example the EF-hand helix-loop-helix pattern analysed by Kretsinger (1980)], DNA binding patterns, and cyclic nucleotide binding patterns amongst others.

Another approach to pattern analysis is to closely examine the sequence preferences for sub-structures at an extreme level of detail. This is the approach that has been followed by Thornton *et al.* (1988 — review) where the turns and loops between β-sheets and α-helices have been rigorously classified into different structural groups, and then analysed for sequence preferences at each position in the resulting patterns.

At the other end of the spectrum, Bashford, Chothia and Lesk (1987) have studied the pattern of sequence preference for a very large-scale sub-structure : the globin fold. Having comprehensively analysed the structural features of the globin family, a pattern matching approach was use to correlate conserved sequence patterns with the known structural roles of each position in the fold. Conservation of amino-acid properties formed the core of the pattern method [comparable to the set method of Taylor (1986)]. The resulting patterns derived from the globin analysis were as might be expected, highly specific to globins managing to accurately discriminate between globins and non-globins when applied to the available NBRF-PIR sequences. This method of laboriously constructing patterns manually from structural studies of protein families contrasts sharply with the automated approaches of for example Taylor (1986). Both methods have both plus and minus points. Automatic procedures are obviously quicker, easier to use, and can work on families for which no structural information is available. Manual procedures allow the problems of multiple alignment that can cause problems for the automatic methods to be circumvented, and produce patterns with a true structural significance.

Recent Work

Rooman and Wodak (1988) and Rooman *et al.* (1990), have recently analysed the predictive power of sequence motifs in an attempt to gain some insight into the current limitations in structure prediction accuracy. Their work involved the generation of short motifs that regularly occur in identical conformations. By extracting such motifs from different sized data-sets they surmise that the current lack of success for pattern based structural prediction is simply due to the limited size of the current structural database. Though within the scope of the study the results are valid enough, it may be unduly pessimistic in that it does not consider the possibilities of combinations of different prediction techniques. For example there is the possibility of using logic based expert knowledge to fill in gaps in the current database [ref. Cohen *et al.* (1983)]. Rooman, Wodak and Thornton (1989) analysed the predictive power of recurrent turn motifs [Thornton *et al.* (1988)] in a similar way to that of Rooman and Wodak's study of automatically generated motifs. The general conclusion of the study was yet again disappointingly pessimistic showing their stand-alone predictive power to be poor. On a positive note the study demonstrates a sound statistically based method for validating consensus patterns, which is an area of pattern generation and matching that is somewhat under-explored.

The 'quality' of patterns is currently an ill-defined parameter. Certainly to date every pattern matching study has considered to a degree the sensitivity and specificity of the pattern under discussion, but few attempts have been made to quantify this information in a rational form. By sensitivity we mean the ability of a pattern to extract all the sequences used in its construction, and by specificity we mean the ability of a pattern to match the constituent sequences *and no others*. Sampling methods as used by Rooman, Wodak and Thornton are just one of a number of possibilities in measuring the effectiveness of generated patterns. Another approach is to consider a pattern as a collection of statistically independent strings of symbols, and to evaluate the expected frequency of occurrence of the total pattern. Hodgman (1989) shows how such probabilities can be calculated for patterns based on amino- acid property sets [e.g. Taylor (1986)].

An attempt at accurately quantifying the statistical significance of automatically generated protein sequence patterns forms an important part of the work by Smith and Smith (1990) on the PLSEARCH package. The automatic pattern methods of Smith and Smith is a somewhat simpler version of the pattern generation scheme of Taylor (1986). In place of a Venn diagram, in this method a strictly hierarchical grouping of amino-acids is used, subdividing the set of amino-acids into small mutually exclusive property sets. Whereas Taylor uses a seed alignment to initiate a cyclic alignment/clustering process, the method of Smith and Smith attempts to boot-strap the pattern creation by aligning each sequence in a clustered family of sequences to a rigid non-consensus pattern string, according to a binary tree ordered by the pairwise similarity score between each sequence and all others. The examples of the PLSEARCH pattern generation algorithm quoted in the paper are all relatively trivial, in that the alignments are fairly clear-cut. How such a rigid pattern generation scheme fairs when presented with highly divergent families of proteins is an interesting question. The danger of misalignment is present when the alignment is performed without the use of consensus patterns. In a sense the method pre-empts the problem by using an insensitive clustering scheme to collect sequences together into families.

Future developments in protein sequence pattern matching will hopefully improve on both the sensitivity of current methods and the speed at which they execute. With over 25000 sequences in the Leeds OWL Database [Bleasby and Wooton (1990)] the slowness of existing approaches is already becoming a problem. Recently a very rapid, and relatively sensitive, databank searching program (BLAST - Basic Local Alignment Search Tool) has been described [Altschul and Lipman (1990)]. Though the speed of BLAST is impressive (searching around 500,000 amino-acid residues per second on a SUN 4/280), and its statistical properties are well characterised [Karlin and Altschul (1990)], it still falls far short of the best known template methods with respect to sensitivity.

The PROSITE database [Bairoch (1989)] is an example of another future research avenue. To provide a useful tool for sequence and structure analysis, pattern match-ing methods must be linked to reliable and comprehensive libraries of pattern data. PROSITE represents a first step in providing such a data resource. PROSITE is a well-documented library of semi- automatically generated protein sequence patterns (of the regular expression class), now integrated into the SWISS-PROT protein sequence database, and distributed by EMBL. Pattern libraries such as PROSITE will almost certainly play an important part in the identification and analysis of the vast number of sequences that will be produced as the various large-scale sequencing efforts begin to bear fruit.

Although only a small fraction of the total field of the creation and application of protein sequence patterns has been presented here, it is hoped that this represents a fair cross-section, and that the potential of this diverse collection of techniques in the vitally important areas of protein structure and function prediction is clearly marked.

References

Abarbanel, R. M., Wieneke, P. R., Mansfield, E., Jaffe, D. A., and Brutlag, D. L. (1984). *Nucleic Acids Res.*, 12:263–280.

Altschul, S. F., Gish, W., Miller, W., Myers, E. W., and Lipman, D. J. (1990). *J. Mol. Biol.*, 214.

Anfinsen, C. B., Haber, E., Sea, M., and White, F. H. (1961). *Proc. Natl. Acad. Sci. USA*, 47:1309–1314.

Bairoch, A. (1989). EMBL Biocomputing Technical document 4, EMBL, Heidelberg.

Barker, W. C. and Dayhoff, M. O. (1972). *Atlas of Protein Sequence and Structure*, 5:101–110.

Bashford, D., Chothia, C., and Lesk, A. M. (1987). *J. Mol. Biol.*, 196:199–216.

Bleasby, A. and Wootton, J. (1990). *Protein Eng.*, 3:153–159.

Blundell, T. L., Sibanda, B. L., Sternberg, M. J. E., and Thornton, J. M. (1987). *Nature*, 326:347–352.

Chou, P. Y. and Fasman, G. D. (1974). *Biochemistry*, 13:212–245.

Cohen, F. E., Abarbanel, R. M., Kuntz, I. D., and Fletterick, R. J. (1983). *Biochemistry*, 22:4894–4904.

Cohen, F. E., Abarbanel, R. M., Kuntz, I. D., and Fletterick, R. J. (1986). *Biochemistry*, 25:266–275.

Cohen, F. E., Sternberg, M. J. E., and Taylor, W. R. (1980). *Nature*, 285:378–382.

Cohen, F. E., Sternberg, M. J. E., and Taylor, W. R. (1982). *J. Mol. Biol.*, 156:821–862.

Dayhoff, M. O., Schwartz, R. M., and Orcutt, B. C. (1978). In Dayhoff, M. O., editor, *Atlas of Protein Sequence and Structure*, pages 345–358. Nat. Biomed. Res. Foundation, Washington, DC. Supplement 3.

Doolittle, R. F., Feng, D. F., Johnson, M. S., and McClure, M. A. (1986). *Cold Spring Harbor Symp. Quant. Biol.*, 51:447–455.

Garnier, J., Osguthorpe, D. J., and Robson, B. (1978). *J. Mol. Biol.*, 120:97–120.

Gribskov, M., Luthy, R., and Eisenberg, D. (1990). *Meth. Enzymol.*, 188:146–159.

Hodgman, T. C. (1989). *Comput. Applic. Biosci.*, 5:1–13.

Karlin, S. and Altschul, S. F. (1990). *Proc. Natl. Acad. Sci. USA*, 87:2264–2268.

Kretsinger, R. H. (1980). *Crit. Rev. Biochem.*, 8:119.

Levitt, M. (1976). *J. Mol. Biol.*, 104:59–107.

Lim, V. I. (1974a). *J. Mol. Biol.*, 88:873–894.

Lim, V. I. (1974b). *J. Mol. Biol.*, 88:857–872.

Lipman, D. J. and Pearson, W. R. (1985). *Science*, 227:1435–1441.

Nagano, K. (1977). *J. Mol. Biol.*, 109:251–257.

Nagano, K. (1980). *J. Mol. Biol.*, 138:797–832.

Needleman, S. B. and Wunsch, C. D. (1970). *J. Mol. Biol.*, 48:443–453.

Palau, J. and Puigdomenech, P. (1974). *J. Mol. Biol.*, 88:457–469.

Pearl, L. H. and Taylor, W. R. (1987). *Nature*, 328:351–354.

Pearson, W. R. (1990). *Meth. Enzymol*, 188:63–98.

Richmond, T. J. and Richards, F. M. (1978). *J. Mol. Biol.*, 119:537–555.

Robson, B. and Osguthorpe, D. J. (1979). *J. Mol. Biol.*, 132:19–51.

Rooman, M. J., Rodriguez, J., and Wodak, S. J. (1990). *J. Mol. Biol.*, 213:337–350.

Rooman, M. J. and Wodak, S. J. (1988). *Nature*, 335:45–49.

Rooman, M. J., Wodak, S. J., and Thornton, J. M. (1989). *Protein Engineering*, 3:23–27.

Schiffer, M. and Edmundson, A. B. (1967). *Biophys. J.*, 7:121–135.

Schiffer, M. and Edmundson, A. B. (1968). *Biophys. J.*, 8:29–39.

Sellers, P. H. (1974). *J. Combinator. Theor.*, 16:253–258.

Sellers, P. H. (1984). *Bull. Math. Biol.*, 46:501–514.

Smith, R. F. and Smith, T. F. (1990). *Proc. Natl. Acad. Sci.*, 87:118–122.

Smith, T. F. and Waterman, M. S. (1981). *Adv. Appl. Math.*, 2:482–489.

Staden, R. (1982). *Nucleic Acids Res.*, 14:363–374.

Taylor, W. R. (1986). *J. Mol. Biol.*, 188:233–258.

Taylor, W. R. (1988a). *Protein Engineering*, 2:77–86.

Taylor, W. R. (1988b). *J. Mol. Evol.*, 28:161–169.

Taylor, W. R. (1990). *Meth. Enzymol.*, 188:456–474.

Taylor, W. R. and Thornton, J. M. (1983). *Nature*, 301:540–542.

Taylor, W. R. and Thornton, J. M. (1984). *J. Mol. Biol.*, 173:487–514.

Thornton, J. M., Sibanda, B. L., Edwards, M. S., and Barlow, D. J. (1988). *BioEssays*, 8:63–69.

Thornton, J. M. and Taylor, W. R. (1989). In Findlay, J. and Geisow, M. J., editors, *Protein Sequencing: A Practical Approach*, chapter 7, pages 147–194. IRL Press, Oxford.

Walker, J. E., Saraste, M., Runswick, W. J., and Gay, N. J. (1982). *EMBO J.*, 1:945–951.

Waterman, M. S. (1986). *Nucleic Acids Res.*, 14:9095–9102.

Wierenga, R. K. and Hol, W. G. J. (1983). *Nature*, 302:842–844.

Wilbur, W. J. and Lipman, D. J. (1983). *Proc. Natl. Acad. Sci.*, 80:726–730.

Wooton, J. C. (1974). *Nature*, 252:542–546.

Multiple Sequence Alignment and Flexible Pattern Matching

Geoffrey J. Barton

Laboratory of Molecular Biophysics
The Rex Richards Building
South Parks Road
Oxford, OX1 3QU
U.K.

geoff@uk.ac.ox.biop

Introduction

Sequence alignments are of central importance in the interpretation of protein sequence data. Alignment techniques may be used to identify previously unsuspected homology, whilst multiply aligned sequences may reveal patterns of residues structurally or functionally important to the family. In this article automated techniques for pairwise and multiple sequence alignment are described together with methods for their evaluation, and improvement by the incorporation of secondary structural information. In addition, the use of *flexible patterns* to describe the important features of a protein fold is explained, and evaluated by comparison to several commonly used alignment procedures.

Background to Automatic Protein Sequence Alignment

The simplest method for the comparison of two sequences considers the sequences as a whole, and does not allow gaps. One protein sequence is slid relative to the other and scores are accumulated for each alternative arrangement. For two sequences of length 100 residues, this operation takes only 100 steps. However, due to the severe limitations of the method, it is rarely used today.

Fixed segment methods (e.g. see Fitch, 1966; McLachlan, 1971, 1972; Argos, 1987) compare all possible overlapping fragments from each sequence (say of length 7 amino acids). The score for the comparison of each fragment pair is either subjected to statistical analysis, or plotted on a contoured display. The method is very powerful for the identification of significant local similarities, but does not directly consider insertions and deletions, nor does it produce a complete alignment of the sequences. Application of fixed segment methods to the 100 residue sequences would require $\simeq 10000$ comparisons to be performed.

Ideally we want to obtain a best alignment of the sequences that includes a consideration of insertions and deletions. The simplistic approach to performing this operation would be to generate all possible alternative alignments, considering all combinations of gaps, obtain a score each alignment, then select the best. However, this systematic method leads to an impossibly large calculation. For example, for two sequences of length 100, there are roughly 10^{75} alternative alignments that would need to be generated.

The technique of dynamic programming which was introduced in molecular biology by Needleman and Wunsch (1970) provides a solution. This method calculates the similarity between two sequences and gives a score for the best possible alignment in $\simeq MN$ steps (where M and N are the lengths of the sequences). Although this algorithm and related methods have been widely described and discussed elsewhere (e.g. see Sankoff and Kruskal, 1983), a brief account is given here so that the multiple alignment, and flexible pattern algorithms may be seen in context.

The Needleman and Wunsch (1970) algorithm may be divided conceptually into three stages. Before applying the method, a scoring scheme must be chosen for aligning individual pairs of amino acids. The simplest scheme is to score 1 for identical pairs and 0 for all other combinations. However, more sophisticated schemes that take into account conservative amino acid replacements, (i.e. similar physico-chemical types), or substitutions observed in protein families, are normally used. Studies by Dayhoff et al.(1978) and Feng et al.(1985) suggest that the Dayhoff mutation data matrix is amongst the best for assessing weak sequence similarities.

1. A matrix is constructed with the two sequences to be aligned, one on each axis. Each cell of the matrix is filled in with the score from the chosen scoring scheme. In Fig. 1a, this is shown in the upper part of the matrix for the simple identity scoring scheme. For example, position 4 of the vertical sequence and 5 of the horizontal sequence are both Ile residues, so the corresponding matrix cell is set to 1. Conversely, position 4 of the vertical sequence is different from position 4 of the horizontal sequence (N), so the intersecting cell is set to 0 (blank in the figure).

2. Once this initial matrix is complete, a process of back-summation is performed to locate the best score for an alignment of the two sequences. This process is shown partially completed in the lower part of Fig. 1a. For example, cell 6,6 is currently being

(a)	1 A	2 W	3 C	4 N	5 I	6 R	7 Q	8 C	9 L	10 C	11 R	12 P	13 M
1 A	1												
2 I					1								
3 C			1					1		1			
4 I					1								
5 N				1									
6 R						1	4	3	3	2	2	0	0
7 C	3	3	4	3	3	3	3	4	3	3	1	0	0
8 K	3	3	3	3	3	3	3	3	3	2	1	0	0
9 C	2	2	3	2	2	2	2	3	2	3	1	0	0
10 R	2	1	1	1	1	2	1	1	1	1	2	0	0
11 B	1	2	1	1	1	1	1	1	1	1	1	0	0
12 P	0	0	0	0	0	0	0	0	0	0	0	1	0

(b)													
1 A	8	7	6	6	5	4	4	3	3	2	1	0	0
2 I	7	7	6	6	6	4	4	3	3	2	1	0	0
3 C	6	6	7	6	5	4	4	4	3	3	1	0	0
4 I	6	6	6	5	6	4	4	3	3	2	1	0	0
5 N	5	5	5	6	5	4	4	3	3	2	1	0	0
6 R	4	4	4	4	4	5	4	3	3	2	2	0	0
7 C	3	3	4	3	3	3	3	4	3	3	1	0	0
8 K	3	3	3	3	3	3	3	3	3	2	1	0	0
9 C	2	2	3	2	2	2	2	3	2	3	1	0	0
10 R	2	1	1	1	1	2	1	1	1	1	2	0	0
11 B	1	2	1	1	1	1	1	1	1	1	1	0	0
12 P	0	0	0	0	0	0	0	0	0	0	0	1	0

FIGURE 1. Operation of the Needleman & Wunsch (1970) algorithm. (a) Match matrix partially completed. (b) Completed matrix showing traceback of paths for alignment. (See text for details).

computed. The score to be stored in this cell is taken as the sum of the value already in the cell (1), and the maximum value of cell 7,7 and the row and column highlighted (4). This gives a value of 5. This operation is performed row by row working from the bottom right hand corner to the top left of the matrix. In the example shown, insertions and deletions (gaps) score zero, however, in general it is desirable to assign a penalty to the introduction of a gap in the alignment. The gap-penalty is incorporated in the algorithm as a number that is subtracted from the row or column score, prior to finding the maximum value. For example, if a gap-penalty of 1 (regardless of the number of residues in the gap) were to be used when computing the score for cell 6,6 of Fig. 1a, we would take the maximum of cell 7,7 and the row and column values each reduced by 1 $(1 + \max(3,4-1,3-1,\ldots 0-1,3-1,2-1,\ldots,0-1) = 4)$.

3. Fig. 1b shows the completed matrix. Each cell of this matrix gives the score for an alignment starting with the alignment of the residues at that cell. For example, cell 9,9 gives the best score for comparing CRBP and LCRPM (2) assuming the initial C and L are aligned. It follows that the best score for the complete alignment of the sequences must be in the first row or column of the matrix and a path may be traced back through the matrix in order to produce an alignment of the two sequences (Fig. 1b). The figure also illustrates that there may be alternative paths through the matrix with the same score.

The dynamic programming technique is guaranteed to give the best possible alignment, but it is only the best alignment in terms of the particular amino acid relatedness matrix used, and gap-penalty parameters chosen. Some of the possible matrices are based on the genetic code or chemical properties, whilst gap-penalties often consist of both a length-independent term (the cost for creating a gap) and a length-dependent term (the cost for extending a gap beyond one residue).

Given a method that can give the best possible alignment for the model, (Dayhoff matrix, plus gap penalty), we want to know whether the model is actually producing biologically meaningful alignments. It is the disposition of the amino acid residues in the native three-dimensional structure of the protein that determines the function. Accordingly, one method of evaluation is to look at families of proteins of known three-dimensional structure, and assuming the structure is conserved more highly than the sequence, use a structural alignment as a measure of the accuracy of the alignment obtained using the sequence information alone. For example, the super-position of immunoglobulin constant domains (Barton and Sternberg, 1988) shows that in certain regions, especially within the core of the protein, (the two β-sheets), the structures are similar, and we can be very confident about the resulting structure-

```
chym  IVNGEEAVPGSWPWQVSLQDKT    GFHFCGGSLINENWVVTAAHCGVT TSDVVVAGEFDQGSSSE KIQKLKIAK
tryp  IVGGYTCGANTVPYQVSLNS     GYHFCGGSLINSQWVVSAAHCYKS GIQVRLGEDNINV VEGNEQFISASK
elas  VVGGTEAQRNSWPSQISLQYRSGSSWAHTCGGTLIRQNWVMTAAHCVDRELTFRVVVGEHNLNQ NNGTEQYVGVQK

SS    SBS EE    TTSSTTEEEEE SS    EEEEEEESSSSEEEE GGG     TTSEEEES  BS  SS  S  EEEEEE
SS    BS EE    TTSSTTEEEEES      SSEEEEEEEEEETTEEEE GGG    S  S EEEES SSTTS    S  EEEEEE
SS    BT EE    TTS TTEEEEEEEEETTEEEEEEEEEEEEETTEEEE GGGG S    EEEEES  BTTS     S  EEEEEE

SCR   xxxxxxx   xxxxxxxxxxx      xxxxxxxxxxxxxxxxxxxxx    xxxxxxxxx        xxxxxxx
```

Kabsch and Sander Summary of Secondary Structure

E/B = Extended conformation
H/G = Helix
T/S = Turn/Bend

FIGURE 2. Part of three serine protease sequences (chym: chymotrypsin, tryp: trypsin, elas: elastase) aligned according to superposition of the three-dimensional structures as shown by Greer (1981). SS lines refer to assignments of secondary structure according to Kabsch and Sander (1983) as shown. SCR identifies Structurally Conserved Regions common to all three proteins as defined by Greer (1981). The insertions and deletions all occur in loop regions joining structurally conserved secondary structures.

derived sequence alignment. In other regions, for example the loops, there is a great deal more structural variation and it is much more difficult to obtain an unambiguous sequence alignment. Indeed, it is often true to say that no biologically or structurally meaningful alignment exists in a particular region. Therefore, for the purposes of testing sequence alignment methods, I choose to look purely at the conserved core of secondary structure regions. It should also be noted that insertions and deletions generally occur between elements of secondary structure. For example Fig. 2 shows part of three serine proteases with the secondary structure assignments according to Kabsch and Sander's program DSSP (Kabsch and Sander, 1983). The insertions and deletions occur between the β-strands and not within the conserved structure. This suggests that one way in which an alignment method might be improved is to reduce the gap penalty in loops relative to these core secondary structures. Of course, this can only be done if the three dimensional structure of at least one of the proteins being aligned is already known. (Using secondary structure prediction as a guide is not very effective because the methods cannot delineate the secondary structures sufficiently well).

If we systematically include secondary structure dependent gap-penalties (Barton and Sternberg, 1987a) and look at the accuracy of the alignments obtained with and without the secondary structural information, we see an improvement for all proteins tested (Fig. 3). It is also clear from Fig. 3 that although three of the sequence pairs are

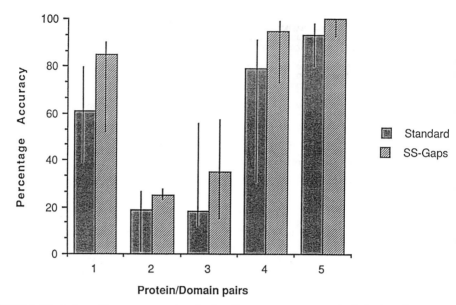

FIGURE 3. Illustration of improvements in pairwise alignment accuracy on the inclusion of a secondary structure dependent gap-penalty for 5 protein pairs. Vertical lines join the best and worst accuracies obtained using 121 different gap-penalty combinations, whilst the solid blocks show the mean values (Barton and Sternberg, 1987ba).

aligned to a very high accuracy, (around 80 per cent), two of them give poor alignments of less than 50% accuracy.

A standard method of assessing the similarity of two sequences is to calculate a normalised score, normally the standard deviation score (z), for the comparison. This is obtained by shuffling the order of the amino acids in each of the sequences and obtaining a value for the comparison of the shuffled sequences. This process is repeated a number of times (typically 100) and the score for the native sequence comparison is expressed in standard deviation units away from the mean of the 100 shuffled sequence values.

The standard deviation scores for the alignments of Fig. 3 show that the accuracy of alignment is reasonably well correlated. As a rough rule of thumb, if the sequence pair scores above six standard deviations from the mean, we can be confident of the alignment accuracy, but below this value more care is needed. This finding agrees with the work of Dayhoff (Dayhoff *et al.*, 1978) who looked at sensitive methods for identifying weak sequence similarity and concluded that six standard deviations from the mean random sample was strong evidence that the two proteins are related.

On a slightly larger data set, Fig. 4 shows the accuracy obtained when comparing

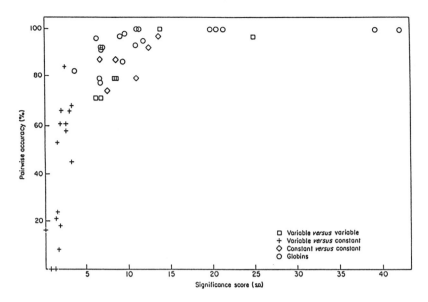

FIGURE 4. Relationship between the accuracy of alignment within the conserved core, and the significance score for pairs of immunoglobulin sequences and globins.

sequences pair-wise within seven globins and eight immunoglobulin domains. For those alignments with high significance (>6 s.d.), the accuracy is very good, while below five to six standard deviations, quality is much more variable. The important feature of this is that with high scores there are *no* bad alignments, so it is possible to predict from the score when the alignment is likely to be good. This observation leads to an approach to multiple sequence alignment, because if we can be confident of the alignment of two sequences, then we should be able to then extend this approach to aligning many sequences.

Multiple Sequence Alignment

In theory it is possible to extend directly the Needleman and Wunsch algorithm to align any number of sequences, however this is not practical due to computer storage and time requirements. A two sequence comparison requires storage proportional to N^2, for three sequences it is N^3 and so on. Successful programmes have been written to align three sequences (e.g. Jue *et al.*, 1980; Murata *et al.*, 1985). In practice however, it may not be particularly useful to use such a rigorous approach to multiple alignment, since for the less closely related sequences, it can be seen from the pair-wise scores

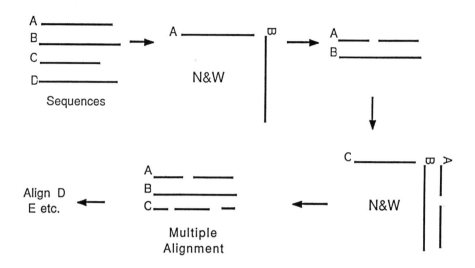

FIGURE 5. Schematic representation of the multiple alignment algorithm. N&W indicates the application of the Needleman and Wunsch (1970) algorithm. The important step is the comparison of sequence C to the pre-existing alignment of sequences A and B. Further sequences may be added by performing the same type of pair-wise comparison.

that any alignment according to a matrix/gap-penalty model is likely to be ambiguous.

My multiple alignment procedure works as follows (Barton and Sternberg, 1987b). I start off by comparing all sequences pair-wise. The most similar pair is then aligned by the Needleman and Wunsch algorithm because that's the pair in which we have greatest confidence of the outcome. The next most similar sequence to the existing alignment of two sequences is then identified and aligned to the existing two sequence alignment, again using Needleman and Wunsch. This is the crucial step, we take an alignment and align another sequence to the existing alignment without changing the existing alignment. Having aligned three sequences, the next closest pair between the three sequence alignment and the sequences yet to be aligned is identified and aligned to the growing family. By adding one more sequence at a time, a complete alignment is built up (see Fig. 5). Each alignment operation is therefore a straightforward pairwise comparison, the only complication is that one of the "sequences" to be aligned is an alignment itself. Accordingly, when obtaining scores to fill the comparison matrix (e.g. Fig. 1a), mean values are taken at each position for the comparison of a single residue and a group of aligned residues (see Barton and Sternberg, 1987b, and Barton, 1990, for full details).

If we compare the accuracy of aligning seven globins and eight immunoglobulin domains by this multiple, and by pair-wise methods, for some sequence pairs there's a dramatic improvement (Fig. 6). However, in general the multiple alignment gives around the same accuracy as the pair-wise method or slightly better. In a few cases the accuracy is worse. If we think about the pair-wise scores in a traditional way using cluster analysis, (which is effectively what the alignment method is doing), and look at globin sequences (Fig. 7), we find that clustering occurs at around six standard deviations and all pairs are aligned very well.

However, if we look at the immunoglobulin domains where we have four variable and four constant domains as a test set (Fig. 7b), it can be seen that, although constant domains cluster at a high score and the variable domains also cluster at a high score, there is a similarity of only $\simeq 3.0$ S.D. between the two groups. We know from our analysis of pair-wise alignments that a pair-wise alignment is likely to be poor at this level, or is at best unpredictable. It is not particularly surprising therefore, that we get some deterioration in the overall alignment quality when we take all Ig domains as a group.

We can conclude from this study that it is important to consider closely related clusters first and then think carefully about how they should be joined. Using this approach it is possible to align very large numbers of sequences (relatively automatically) with a high degree of confidence, and also in a reasonable amount of computer time. The flexible pattern matching approach discussed in the following section provides one method of assisting in the alignment of such distantly related clusters of sequences.

Flexible Pattern Matching

One approach to express and locate patterns in protein sequences is to use a regular expression pattern language (e.g. Abarbanel *et al.*, 1984) . An alternative, is to use a fixed length segment and a weight matrix (e.g. Dodd and Egan, 1987). And yet another approach is using a series of templates or profiles (e.g. Taylor, 1986, Bashford *et al.*, 1987). I will define a *flexible pattern* (Barton and Sternberg, 1990; Barton, 1990) as a series of elements or segments, perhaps made up of a section from an alignment, something we know from structure (e.g a conserved helix) or a particular required residue (from biochemical information) separated by gaps of a defined length range (Fig. 8).

In principal, a very rich picture of the protein of interest can be built up using flexible

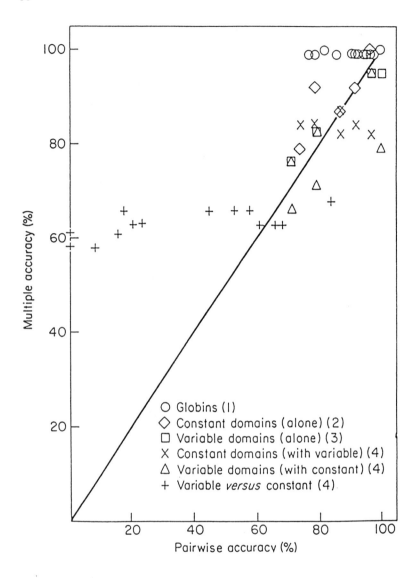

FIGURE 6. Comparison of accuracy obtained within multiple alignments by the algorithm, when compared to pairwise alignments. Points above the line indicate an improvement in alignment accuracy.

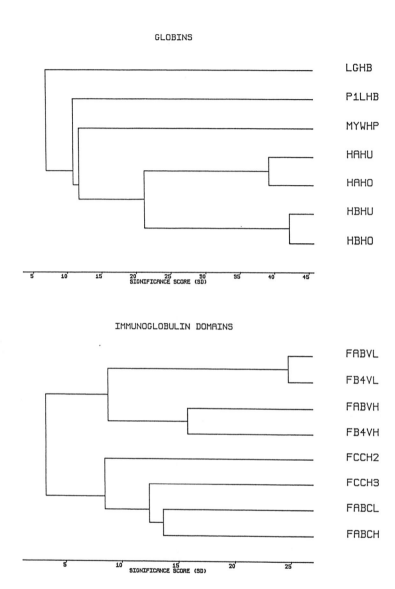

FIGURE 7. Dendrograms for the pairwise comparison of (a) 7 globin sequences and (b) 8 immunoglobulins. The point at which all sequences cluster suggests the expected accuracy of alignment for the family.

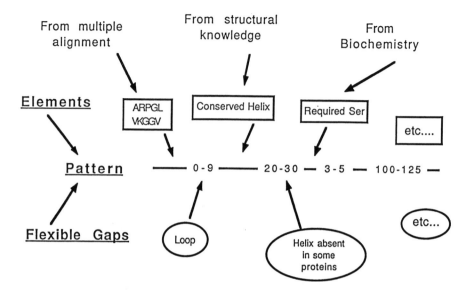

FIGURE 8. A hypothetical flexible pattern that incorporates information from a variety of sources to provide a description of the protein under study.

patterns. The flexible pattern method combines both the weight matrix approach, some aspects of regular expression matching and the concept of variable length gaps. If we represent the pattern as an alignment, we can use any sort of weighting scheme. In Fig. 9 we have a very simple example, using identity scoring (a score of one for match and zero for non-match). But, of course, more sophisticated schemes can be used, for example averaged scores of the Dayhoff matrix, as were used in the multiple alignment method.

How do we find the best alignment of a pattern with a particular sequence, in what is potentially a very large search space? This is a similar problem to the two-sequence alignment comparison, and a dynamic programming algorithm provides a solution. Initially, a matrix is filled with scores for each segment of the pattern versus each amino acid, then we work back through the matrix, as with the Needleman and Wunsch method, except that in this case instead of considering all possible gap lengths we simply look within the region that the gap is allowed (see Fig. 10).

Having completed this process we find the highest scoring cell in the first column and trace back to obtain the alignment of the pattern and the sequence. We can also look for repeats of the same pattern, and alternative tracings through the matrix, and so on.

Pattern Segments (E_i)		Gap Lengths (F_i) Min | Max		Lookup Table $(T_{N, 23})$ (Partial)								
				A	R	G	L	P	E	D	W	F ...
1	A A			1.0	0	0	0	0	0	0	0	0
		0	2									
2	R G			0	1.0	1.0	0	0	0	0	0	0
		0	1									
3	P A			1.0	0	0	0	1.0	0	0	0	0
		0	5									
4	D E			0	0	0	0	0	1.0	1.0	0	0
		0	0									
5	W F			0	0	0	0	0	0	0	1.0	1.0

FIGURE 9. A simple flexible pattern showing the lookup table and defined gap lengths. Five pattern segments are shown, each of which corresponds to a position in a sequence alignment. Gap lengths are defined to have minimum and maximum bounds between each segment of the pattern. The table shows the score associated with matching each pattern segment against each possible amino acid. For clarity, simple identity scoring is shown, however, in general a more sophisticated scheme would be used.

(a)

min:max	0:2	0:1	0:5	0:0	Gaps (F_j)
	A A	R G	P A	D E	W F	Segments (E_i)
V			2			1
I			2			2
A	1		3			3
G		4	2			4
T		3	2			5
P		1	3			6
E		1	1	2		7
F		1	1		1	8
S			1			9
D				1		10
	1	2	3	4	5	

(b)

V	3	3	2		
I	3	3	2		
A	5	2	3		
G	3	4	2		
T	1	3	2		
P	1	1	3		
E	1	1	1	2	
F		1	1		1
S			1		
D				1	

FIGURE 10. Illustration of the calculation of the best match between the pattern shown in Fig. 9 and a short sequence. (a) The pattern is shown on the horizontal axis of the matrix whilst the sequence is on the vertical. The matrix is shown partially completed, cell 4,2 has just been processed by taking the highest score in the preceding column, within the defined gap-range of 0:1 and adding this to the score already in cell 4,2. (compare this to Fig. 1a for the Needleman and Wunsch algorithm). (b) The completed match matrix. The highest value in the first column of the matrix shows the start of the best match between the pattern and the sequence — lower scores give partial matches. The alignment of the pattern and the sequence may be traced out as shown. Scores in the first row of the matrix indicate partial matches of the pattern to the sequence.

Of course, as with any technique, we need a good test case for the method. This allows us to determine whether if we define a pattern which, for example, describes a particular protein fold, we get a better result than using a conventional alignment method of some sort. As a test set 345 complete globin sequences were used. These have a similar fold over many species and there have also been detailed studies in the literature (e.g. see Bashford *et al.*, 1987, and refs therein) that provide additional sources of information.

Given a pair of sequences, several methods of comparison may be used. (a) We can use a conventional Needleman and Wunsch approach, which finds the global alignment of the sequences (i.e. the best overall alignment). (b) If we know the three-dimensional structure of this protein, and therefore know where the secondary structural regions are; we can include secondary structure dependent gap penalties, which we would expect to give better results. (c) If more than two sequences are available, multiple alignment methods (possibly aided by the structural criteria) can be used to align the sequences.

When deriving a flexible pattern from an alignment of sequences, the most highly conserved parts of the protein family (the conserved secondary structures), can be used to make a pattern which can then be compared to the database. Between these conserved regions we can look at alternative flexibility schemes, either constrained flexiblity between the minimum and maximum bounds, or perhaps (as other workers have used) an unconstrained gap length.

The approach adopted for testing these methods using the globin family is as follows. We know there are 345 globins in the database so we compare our pattern or sequence to the whole database to obtain a distribution of scores. For the purposes of discussion, three values are determined from the score distribution. 1. how many sequences we find before the first non-globin, 2. the number of globins left in the top 500 sequences, and 3. how many globins are not in the top 500 sequences. The first figure gives an indication of the specificity of the method, whilst the second and third figures show the profile of the scores; a high proportion of globins in group 2 shows a sensitive method with poor specificity whilst a high proportion in group 3 shows poor specificity and sensitivity.This is a fairly arbitrary division, but it gives a feel for the distribution of scores without having to actually show every value.

Fig. 11 shows a section of the multiple alignment of seven globins produced by Bashford *et al.*(1987) based upon three-dimensional structural criteria. Various flexible patterns derived from the alignment are also shown. In Pattern 1, each of the regions shown is a conserved helix separated from its neighbours by a flexible gap. For comparison purposes, we can also use a single sequence from this set to derive a pattern, or for a conventional sequence scan. Since we known the three dimensional structure of the query protein(s), we also have the option of using secondary structure dependent gap-penalties, and since several sequences are available, we may use the entire alignment to scan the database using the multiple sequence comparison method.

Comparing these methods in Table 1, we find the single sequence (human α haemoglobin) scan using the standard database scanning programme FASTP (Lipman and Pearson, 1985), finds 297 globins before the first non-globin, with seven still in the top 500 and 41 missed, which is not a terribly good result. If we use the Needleman and Wunsch method, we do slightly better but not dramatically so (Scan 2). Similarly, if we include secondary structure dependent penalties, we do slightly better still (Scan 3). Taking the structure based alignment of seven globins, we can scan the entire alignment against the database and this produces reasonably good results (Scan 4). Similarly if we include secondary structure penalties in conjunction with that alignment, we do slightly better still (Scan 5). However, the most striking effect is when we just take the secondary structural regions and ignore the more variable loop regions and scan with a pattern. This finds all 345 globins with no misses at all (Scan 6).

In order to examine why the pattern performs so well, I took just one sequence from the set (human α haemoglobin) and made a flexible pattern based upon its secondary structures. This pattern performs almost as well (Scan 7), suggesting that much of the benefit from the pattern method is coming purely from throwing out the loops, the more variable parts of the sequence, and keeping the more highly conserved parts. The method does significantly better than any of the single sequence approaches including the variable gap penalty approach. If we then take two sequences we get almost up to the level of the seven sequence pattern. So really, the main factor governing the improvement, is the discarding of the variable loop regions.

Of course, all these patterns were derived from a very high quality alignment based upon structural super-position, but what happens if we derive a pattern from a multiple alignment of the same seven globins done entirely automatically, and scan with that. The first attempt at defining a pattern automatically (it is very difficult to be unbiased in this) did quite well (Scan 9) even though there are a few errors in the alignment and the rules may not be perfect for deriving the pattern. So we do not need to know the tertiary structure to be able to get a very effective pattern for globin sequences.

It should be noted that all the techniques described so far find the best overall, or *global* alignment of two sequences, or of a pattern and a sequence. Since the

```
                                                  Flexible Patterns
Acc   Structure        Alignment             1   2   3   4   5   6   7   8

                   D D H G G K G
                   L L L   L
                   S S K K T G
      D 1          T T D T T
      D 2          P E L A S                       Flexible Gap
      D 3          D A E D E A
      D 4          A E S Q V S
      D 5          V M I L P
      D 6          H M K K K Q
      D 7          G G A G K N
 e    E 1          S N S T S N D       |       |   |   |   |
 e    E 2          A P E A A P P       |
 e    E 3          Q K D P D E A       |
 b    E 4          V V L F V L V       |       |   |   |   |   |
 e    E 5          K K K E R Q A       |
 e    E 6          G A K T W A D       |
      E 7          H H H H H H L       |
 b    E 8          G G G A A A G       |       |       |   |   |   |
 e    E 9          K K V N E G A       |
 e    E10          K K T R R K K       |       |           |   |   |
 b    E11          V V V I I V V       |       |       |   |   |
 b    E12          A L L V I F L       |
 e    E13          D G T G N K A       |
 b    E14          A A A F A L Z       |
 b    E15          L F L F V V I       |       |           |   |   |
      E16          T S G S N Y G       |           |   |
 e    E17          N D A K D E V       |
 b    E18          A G I I A A A       |       |       |   |   |
 b    E19          V L L I V A V       |
 e    E20          A A K G A I S       |
      EF1          H H K E S Q H       |
                   V L   L M L L
                   E
                   V
                       P D T G
                   D D K   D G D
                       T V Z                       Flexible Gap
                     G   E V G
                   D N H N K V K
                   M L H I M S M
                   P K E E S D V
                   N G A A M A A
                   A T E D K T Q
 e    F 1          L F L V L L M       |       |   |   |   |
 e    F 2          S A K N R K K       |
 e    F 3          A T P T D N A       |
 b    F 4          L L L F L L V       |       |   |   |
 b    F 5          S S A V S G G       |
 e    F 6          D E Q A G S V       |
      F 7          L L S S K V R       |
      F 8          H H H H H H H       |       |   |   |   |   |   |
      F 9          A C A K A V K       |
      F10          H D T P K S G       |
 e    FG1          K K K R S K Y       |       |
      FG2          L L H G F G G       |
```

FIGURE 11. Part of the structure-based alignment of seven globins reported by Bashford *et al.*(1987). Acc: accessibility of the residues in the folded protein following Bashford *et al.*(1987) — e = exposed, b = buried. Structure: The residues participating in helix D, E and F are labelled. Flexible Patterns: Vertical bars show the positions of pattern elements selected from the multiple alignment. As we progress from pattern 1 to pattern 8 successively fewer, yet more highly conserved positions make up the pattern.

Database Scans Using Queries Derived From Globin Sequences						
Scan Number	Source of Query	Method (Gap-Penalty)	Additional Structural Information?	Globins Before First Non-Globin	Globins Remaining in Top 500 Scores	Globins not in Top 500 Scores
1	Single Sequence (HAHU)	FASTP	No	297	7	41
2		NW(16)	No	306	8	31
3		NW_SS(16)	Yes	311	9	25
4	7 Globins (3D Structure Alignment)	BS(16)	No	309	19	17
5		BS_SS(16)	Yes	318	12	15
6		FP	Yes	345	0	0
7	Single Sequence (HAHU)	FP	Yes	337	7	1
8	Two Sequences (HAHU, GGICE3)	FP	Yes	344	1	0
9	7 Globins (Automatic Multiple Alignment)	FP	No	327	18	0

TABLE 1. The result of database scans using queries derived from globin sequences. Source of query: Protein Identification Resource (PIR) codes are shown for the sequences used, 3D Structure alignment refers to the alignment reported in Bashford *et al.*(1987). Method: FASTP = FASTP programme (Lipman and Pearson, 1985), NW = conventional Needleman and Wunsch algorithm, NW_SS = NW with secondary structure dependent gap-penalties (i.e. reduced penalty for gaps in loop regions), BS = multiple alignment following Barton and Sternberg (1987b), BS_SS = BS with secondary structure dependent penalties, FP = Flexible Pattern method. Additional Structural Information: signals Yes if information in addition to the sequence/alignment is utilized for the scan (e.g. positions of α-helices from X-ray crystallography). Globins Before First Non-Globin: the number of globin sequences that give scores higher than a non-globin sequence. Globins Remaining in top 500 Sequences: Globins scoring less than a non-globin, but still in the top 500 sequences.

Comparison to Local Methods

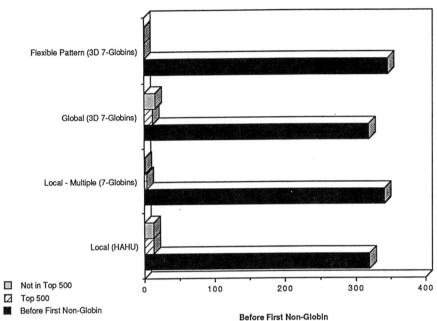

FIGURE 12. Comparison of Global and Local alignment methods using the globin family test system. The top two scans correspond to scans 6 and 2 from Table 1. The third scan (Local - Multiple(7-Globins)) scanned the databank with the 7-globin alignment using the local similarity Smith-Waterman (1981) algorithm, whilst the fourth scan (Local HAHU) shows the result of the same algorithm applied to the Human α haemoglobin sequence. The Local-Multiple method which is similar to the "Profile" approach of Gribskov *et al.*(1987) performs nearly as well as the flexible pattern for this family.

similarities between the sequences are often non-uniform, it may be better to use a technique that looks for the best locally similar regions such as the Smith and Waterman (1981) algorithm. This method is a modified Needleman and Wunsch method that locates the best local similarities between two sequences. If we use the method with α haemoglobin, it performs significantly better than the global alignment method, similarly, applying the Smith-Waterman algorithm to the comparison of an alignment and a single sequence gives results nearly as good as the flexible pattern method (Fig. 12)

What then is the advantage in using flexible patterns? One thing that they do is to allow great specificity in defining a weighting scheme, and in defining precisely

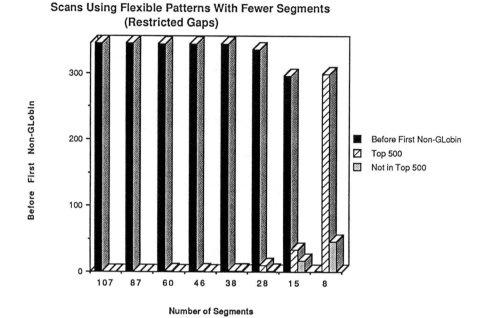

FIGURE 13. The result of reducing the number of elements in patterns derived from the 7-globin alignment of Bashford *et al.*, (see Fig. 11). A serious deteriation in specificity and sensitivity only occurs when the pattern is reduced to 15 elements.

which residue positions are to be involved in the comparison. Accordingly, we can use the technique to explore which are the important residues in a particluar protein fold, by defining patterns that utilise different combinations of highly conserved residue positions. For example, with each of the patterns shown in Fig. 11, progressively more highly conserved residues are included, so in the eighth pattern there are only the very highly conserved positions defined (the ninth pattern would have only those positions that are identical in all 7 sequences). When we scan with each of these patterns and look at the number of globins we find before the first non-globin, in the top 500 and those that we have missed completely we see that as the total number of pattern segments is reduced from 107 (where we have all the secondary structural regions), a very high degree of selectivity for the globins is maintained, down to between 28 and 38 segments. Not until we get down to around 20 per cent of the sequence does the selectivity seriously deteriorate. By pattern eight, selectivity for globins is completely lost, though many of the family still give relatively high scores (Fig. 13).

These results used a pattern with restricted gap lengths between each pattern element, (gaps within a particular range). If we make the gap between secondary structural

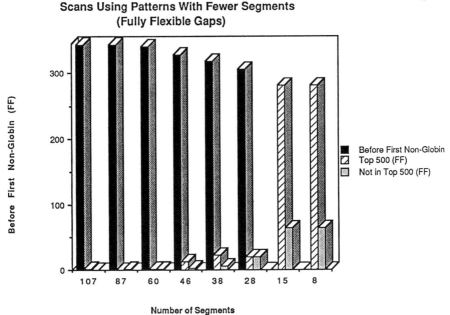

Scans Using Patterns With Fewer Segments
(Fully Flexible Gaps)

FIGURE 14. As for Fig. 13, however, no restriction was placed on the size of the gaps allowed between defined secondary structure pattern elements. The pattern specificity and sensitivity is reduced for patterns containing as many as 46 pattern elements.

regions totally flexible, the selectivity clearly deteriorates much more quickly (Fig. 14). This behaviour is not really suprising since there will be many more opportunities for spurious matching.

All the patterns described and shown in part in Fig. 11 were based purely upon inspection of the 7-sequence multiple alignment. In contrast, Bashford *et al.*(1987) defined templates based upon the positions that were buried or exposed or a combination of those two based on an examination of the three-dimensional structures. The question is, do the patterns used here have any relevance to the three-dimensional structure?

Fig. 15 shows the relationship between the number of pattern segments, and the environment of the residues involved. As the less highly conserved positions are excluded, we find that the buried positions are maintained. Most of the elements in the final patterns are buried hydrophobics, with a few other important highly conserved residues maintained. It is interesting to note that if you just take the buried positions and derive a pattern, the results are not as good as including the highly conserved yet exposed positions. This implies that some of these exposed positions are important to the fold or function of the globins.

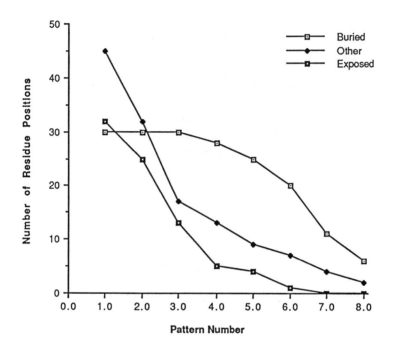

FIGURE 15. Illustration of the structural environment of the pattern elements making up patterns 1–8 from Fig. 11. Exposed residue positions are selectively removed from the progressively sparser patterns. Buried residue positions corresponding to hydrophobic amino acids in the conserved core of the protein are maintained.

Acknowledgements: I thank Dr. M. Sternberg for his guidance, and Drs. C. Rawlings and J. Fox for their encouragement. This article was directly adapted from a transcript of a lecture given at the EMBO workshop on Patterns in Protein Sequence and Structure, EMBL, Heidleberg, September, 1989.

References

Abarbanel, R. M., Wieneke, P. R., Mansfield, E., Jaffe, D. A., and Brutlag, D. L. (1984). *Nucl. Acids Res.*, 12:263–280.

Argos, P. (1987). *J. Mol. Biol.*, 193:385–396.

Barton, G. J. (1990). *Methods Enzymol.*, 183:403–428.

Barton, G. J. and Sternberg, M. J. E. (1987a). *Prot. Eng.*, 1:89–94.

Barton, G. J. and Sternberg, M. J. E. (1987b). *J. Mol. Biol.*, 198:327–337.

Barton, G. J. and Sternberg, M. J. E. (1988). *J. Mol. Graph.*, 6:190–196.

Barton, G. J. and Sternberg, M. J. E. (1990). *J. Mol. Biol.*, 212:389–402.

Bashford, D., Chothia, C., and Lesk, A. M. (1987). *J. Mol. Biol.*, 196:199–216.

Dayhoff, M. O., Schwartz, R. M., and Orcutt, B. C. (1978). In Dayhoff, M. O., editor, *Atlas of Protein Sequence and Structure*, pages 345–358. Nat. Biomed. Res. Foundation, Washington, DC. Supplement 3.

Dodd, I. B. and Egan, J. B. (1987). *J. Mol. Biol*, 194:557–564.

Feng, D. F., Johnson, M. S., and Doolittle, R. F. (1985). *J. Mol. Evol.*, 21:112–125.

Fitch, W. M. (1966). *J. Mol. Biol.*, 16:9–16.

Greer, J. (1981). *J. Mol. Biol.*, 153:1027–1042.

Gribskov, M., McLachlan, A. D., and Eisenberg, D. (1987). *Proc. Natl. Acad. Sci. USA*, 84:4355–4358.

Jue, R. A., Woodbury, N. W., and Doolittle, R. F. (1980). *J. Mol. Evol.*, 15:129–148.

Kabsch, W. and Sander, C. (1983). *Biopolymers*, 22:2577–2637.

Lipman, D. J. and Pearson, W. R. (1985). *Science*, 227:1435–1441.

McLachlan, A. D. (1971). *J. Mol. Biol.*, 61:409–421.

McLachlan, A. D. (1972). *J. Mol. Biol.*, 64:417–437.

Murata, M., Richardson, J. S., , and Sussman, J. L. (1985). *Proc. Natl. Acad. Sci, USA*, 82:3073–3077.

Needleman, S. B. and Wunsch, C. D. (1970). *J. Mol. Biol.*, 48:443–453.

Sankoff, D. and Kruskal, J. B. (1983). *Time Warps, String Edits and Macromolecules — The Theory and Practice of Sequence Comparison.* Addison-Wesley.

Smith, T. F. and Waterman, M. S. (1981). *J. Mol. Biol.*, 147:195–197.

Taylor, W. R. (1986). *J. Mol. Biol.*, 188:233–258.

Discussion

Q: How do you decide the order in which to align the sequences?

A: Based upon how similar each sequence is to the set. This is effectively a simple clustering approach.

Q: Is the clustering based upon the growing multiple alignment, or simply on the pairwise similarity scores?

A: Simply from the pairwise scores.

Q: Why do you use single linkage cluster analysis?

A: Purely because it was the simplest method to code. It also seems adequate for the use here, however, in general I agree that it is probably better to use a more sophisticated method.

T. Blundell: At one point you said your method failed but you got some things on the righthand side of your curve which were less successful. I wasn't sure whether you said later on that if you'd have taken it and, in fact, pair-wise followed them according them to the phylogenetic tree; in other words, taken similarities and come into the root of the tree as Johnson and Doolittle do, whether in fact it would have been successful. And whether, following what Cyrus (Chothia) said, if you'd have clustered in the two groups in the bacterial and the mammalian serine proteases, you could then get a match.

A: I can't comment on the serine proteases because I haven't done that alignment, but certainly with the immunoglobulin example I showed you, it actually makes very little difference whether you follow the entire tree or whether you just take a simple order. The crucial point is what the overall similarity is. If the similarity is high, then you will get a good alignment. If its low, then the quality is inevitably much less predictable.

T. Blundell: It seems to me logical that if things diverge to the number of branch points, the logical way is to go back to each branch point and so end up at the base of the tree. Any other way seems to be almost defying nature.

A: I agree that the tree-based alignment approach is certainly cleaner and intuitively would appear better, but in practice it can make very little difference to the overall alignment when the sequence similarity is high.

Q: Do you find that sparse patterns are generally better?

A: Well, I wouldn't say that the sparse pattern for the globins was good (gives good discrimination). At pattern eight you're actually missing proteins that are really quite similar to the sequences that you've got in your original set. As the number of elements are reduced in the pattern, the first globins that you start to miss are ones which Cyrus (Chothia) reported as having dubious sequences. In addition, there are other globins which I think may have interesting substitutions in the sequences. And then suddenly you start to miss out whole familes, including globins that are part of the original set.

Q: How do you decide on your similarity score? Could it affect when you first encounter a non-member of the family in the list of hits? For example, you could use the Fast P program which has its own similarity score, so is your comparison really objective?

A: No, I agree. However, I have a lot of data on using different similarity schemes which is not shown. The FASTP programme looks at identities first, whereas all the other approaches I described use the Dayhoff matrix, either averaged or directly, so after FASTP they're all consistent with each other. In fact, if you look at specific weight matrices — if you derive a frequency matrix normalised by the observed amino acid frequencies, you don't actually do any better than you do with Dayhoff's matrix — for my data anyway. FASTP is included in the table to show the limitations of what is a very widely used databank searching programme.

Q: Fast P uses one approach, to try to extend the the two scores and I assume you would use some kind of count depending on whether you were hitting the pattern or not?

A: No. With my approach you have a continuous set of scores including all sequences in the database. There is no cutoff, the choice of looking at the top 500 scores is purely to simplify presentation, the figures I showed include scores for sequences "not in top 500" as well.

Q: How important are alternative alignment orders to the final multiple alignment result?

A: Again, that depends upon the overall similarity between all pairs in the set. If they're all fairly similar — if they all cluster pair-wise to well over six standard deviations, then it doesn't really matter what order you do the alignment in. You do get some small variation in the alignments around gaps which you would expect, but the order doesn't matter. If you've got less similar sequences in your set then, again, the order has an effect — you'll get different alignments and you'll tend to get more different alignments because there's less certainty in any single one, so you get a bigger distribution of alternatives.

Q: Do you find that the alignments of the pattern with the sequence are displaced, say by one turn of a helix?

A: Yes, sometimes, so if you were using the alignment as the basis of a modelling study it would be wise to look at several alternative alignments of the pattern and sequence, and assess them each in terms of the predicted structure and other available data.

W. Taylor: Is it possible that the globins were not a good example because they seemed to be too easy to align? If you took the immunoglobin super family where the multiple alignment method failed, I think then your patterns would probably have made an improvement.

A: Yes, I agree, the globins are a relatively straightforward family to align, however the improvements are shown relative to other alignment methods. The immunoglobulin superfamily would certainly be a challenging test of the pattern method and I intend to derive flexible patterns for this family in the near future.

C. Chothia: The important point which you brought up was that these gaps between the secondary structure motifs don't just represent areas where the sequences are very bad. They represent regions which have quite different folds and therefore the lack of sequence homology is real in the sense that their codes are different — they have different structures. If you look at the immunoglobulin domains, and compare the constant domains and variable domains, there's only about 35 residues which have the same fold in the two forms. That's only about one-third or one quarter — one third of the constant domains have the same structure as one quarter of the variable domains and the rest of the protein does not just have a different sequence, it has a different structure. So if you have a sequence alignment in those regions it is not biologically sensible. You can only meaningfully have an alignment for a small part of the protein, so what you are doing for the globins could also be done with variable domains and constant domains, but you'd only expect to find a meaningful result for a small part and that would be correct — there's no way you can align the rest of the sequences. For the immunoglobulins you can align the disulphides, this is part of the one-third of the structure which is the same in both. Two-thirds are different and therefore you cannot align those sequences in a meaningful way; and so it's a real result not to find an alignment.

SCRUTINEER: a Program to Explore and Evaluate Protein Sequence Patterns in Databases

Peter R. Sibbald and Patrick Argos

European Molecular Biology Laboratory
Postfach 10 22 09
Meyerhofstrasse 1
6900 Heidelberg
Federal Republic of Germany

sibbald@EMBL.bitnet

Description of the Program

SCRUTINEER is a computer program designed to search protein sequence databases for features of interest to molecular biologists. The program has been described in detail elsewhere (Sibbald and Argos, 1990a); only a brief description is given here. As originally conceived, the program was required to effect interactive exploration of protein sequences and describe features of interest once located. To enhance portability of the program, it was written in standard Pascal. SCRUTINEER only works with the sequence part of databases and not the annotation since many other programs already search the latter [e.g. PSQ (Sidman *et al.*, 1988), GCG (Devereux *et al.*, 1984)].

Protein sequences can be queried for very complex patterns or targets as (1) multiple amino acids at a position; (2) segments of variable length; (3) segments with delocalized constraints such as composition; (4) segments whose propensity for a particular secondary structure lies within a specified range and (5) constraints of the form "if amino acid 3 is one of set A then amino acid 6 must be one of set B". The search can be confined to particular regions of each sequence; e.g. from amino acid 50 to an amino acid that is 65% of the way along the sequence.

Since searches often find many hits, it is important to accelerate the rate at which the hits could be screened by the user. The program was thus designed to describe what was found, and to allow flexible use of the hits. For example, the distances between hits in the same sequence but from two different searches can easily be listed. The possiblity of combining hits from different searches using boolean operators was also included. The result is a program that is user friendly and reasonably quick while being flexible enough to allow targets of essentially unlimited complexity.

New Features of SCRUTINEER

SCRUTINEER is now provided with a news letter that details modifications, extensions and corrections. An important update is the existence of two program versions: VAX and Standard. While the original criterion of portability argued against a VAX version, many useful features can be incorporated in the latter. Both versions now have self-documenting targets which appear under the DISPLAY, HISTORY part of the program. The number of hits is reported before the search is named and the DISPLAY/OUTPUT now will output the latest hits by default (type 0).

A new weighting algorithm for alignments has been added and is described in detail in Sibbald and Argos (1990b). Briefly, alignments often consist of sequences that are obviously not independent and which should not all receive equal weighting. Such an alignment might contain 4 primate sequence and one nematode sequence. Intuitively this alignment should be weighted so that the 4 primate sequences share a total weight of 0.5 and the nematode sequence is weighted 0.5 (see also Vingron and Argos, 1989 and Altschul et al., 1989 for further discussions of the problem). Of course for larger, more complex problems intuition fails and an algorithmic solution is required. This new weighting algorithm makes searches with alignments more accurate; the incidence of both false positives and negatives is reduced for typical profile (Gribskov et al., 1987) and pattern searches.

In the VAX version, it is now possible to select from several databases within the program and to name the output file from within the program. Protein sequence databases overlap considerably. In order for a search to be complete it should be performed on all databases but with the overlap between databases eliminated for efficiency. Solutions to this problem have been proposed (Claverie and Bricault, 1986; Bleasby and Wooton, 1990). The solution currently implemented at EMBL exploits the PirOnly database (Rice, 1990) which contains the sequences that are in PIR (Sidman et al., 1988) and not in SwissProt (Bairoch, 1990a). From within SCRUTINEER (VAX version), the database choices include (1) SwissProt; (2) OldPIR; (3) NewPIR; (4) AllPIR = OldPIR + NewPIR; (5) PirOnly; (6) Both = SwissProt + PirOnly; or (7) User's own database. With the 'Both' option the combined database (SwissProt release 14.0 + PirOnly derived from PIR release 23.0) consists of 19,311 sequences containing 5.7×10^6 amino acids and with minimal sequence duplication.

A Sample Session

Subsequently is presented a session showing SCRUTINEER searching for the helix loop helix motif described later in the paper. This is the VAX version. Commands typed by the user are in boldface type.

$ **prep scrutineer**

$ **scrutineer**

```
logo appears here


************************ NEW **************************
Scrutineer now allows the following protein sequences
databases or combinations to be used:

   SwissProt = the entire SwissProt database
   Oldpir    = the old part of the PIR/NBRF database
   Newpir    = the new part of the PIR/NBRF database
   Allpir    = Oldpir AND Newpir
   PirOnly   = the sequences in PIR that are NOT in SwissProt
               ( courtesy of Peter Rice )
   Both      = SwissProt AND PirOnly
   User      = the user's own database from their own directory


To use your own database you must have previously created it.
Just type User and you will be asked what its name is.


To use SwissProt, type SwissProt. To use all of PIR type Allpir.

Database choice?
```
both
```
loading command?
```
load
```
loading data... please wait
finished loading data...
top level command?
```
search
```
search command?
```
flex
```
flex search command
```
target

```
please enter flex target now. UPPERCASE LETTERS!
```
[RLMS][EA][RK]{0 1}[RKQI]@[EKQNARS]@@[NKRL]$$[FL]
```
enter more target or hit return when finished
```
@@L$@@@{7 24}[KE]@$@@@@@@@[YVFCH]@$$$
```
enter more target or hit return when finished

a frame with a minimum length of 0
and a maximum length of 1 can be
further modified. type 5 to get help. type 4 when/if you
are done modifying this frame.

number please: 5=help 4=done
```
5
```
Choose from:
1. position constraint e.g. pos 3 of frame is [RK]
2. composition constraint e.g. there are at least 2
   positions that are in [AGLIV]
3. interactive constraint e.g. if pos 4 of frame is in [AG]
   then pos 2 must be F
4. finished adding constraints
5. help (to get an example)
6. parametric constraint - use a mathematical function and
   numerical values for each amino acid. Then specify that
   the function score must be within a desired range for a hit.

Type a number from 1 to 6. Scrutineer will help perform steps.

number please: 5=help 4=done
```
1
```
which position to constrain?
```
1
```
type in BLOCKLETTERS the amino acids that can occur here
```
$
```
type the new amino acids for$
type them now, all on one line in BLOCKLETTERS:
```
ABCDEFHIKLMNPQRSTVWYZ

```
$ = ABCDEFHIKLMNPQRSTVWYZ
is this ok? y/n
```
y
```
number please: 5=help 4=done
```

4

```
type the new amino acids for@
type them now, all on one line in BLOCKLETTERS:
```
ABCDEFHIKLMNQRSTVWYZ

```
@ = ABCDEFHIKLMNQRSTVWYZ
is this ok? y/n
```
y

```
a frame with a minimum length of 7
and a maximum length of 24 can be
further modified. type 5 to get help. type 4 when/if you
are done modifying this frame.

number please: 5=help 4=done
```
4

```
flex search command
```
go

```
number of hits : 50
number of entries inspected : 18185
The search being performed has no name. Type name now.

name must be no more than 60 letters.   60 is here
```
helix loop helix / Rop motif

```
top level command?
```
display

```
display command?
```
hits

```
code       name of search
1 helix loop helix / Rop motif


hits from those entries that have a key hit
will be processed. OR type 0 (zero) to process
latest hits *** NEW ***

type code of key hit
```
0

```
>>>>ARLC$MAIZE
ANTHOCYANIN REGULATORY LC PROTEIN (GENE NAME: LC). 1/90
 420 SERKRREKLNEMFLVLKSLLPSIHRVNKASILAETIAYLKEL 461
```

```
.
other hits...
.
>>>>S03325
Transforming protein (hcr-myc) - Woodchuck
 388 LERQRRNELKRSFFALRDQIPELENNEKAPKVIILKKATAYILSV 432

display command?
pop
top level command?
quit
sure you want to quit? y/n
y
```

And the session ends. Loading the data took 2 min and the entire session took 4 min.

Planned Enhancements

The development of SCRUTINEER continues in response to pressure from users as well as the authors for extended capabilities. A modification of the VAX version currently in progress involves the use of map sections to place the data in virtual memory with minimal load time. As described in detail elsewhere (Sibbald and Argos, 1990a), SCRUTINEER obtains many of its desirable properties by having the data in virtual memory. This provides not only speed but also the possibility of flagging hits for subsequent processing. Unfortunately it takes nearly 3 min on a VAX 8650 to read the data (SwissProt and PirOnly) into virtual memory. With the data as map sections this time will be reduced to a few seconds. It will also allow the 'swapping' in and out of large sections of data as each one is processed and in this way, even when the databases have grown considerably, SCRUTINEER will still be viable.

A second development is searching a new sequence against a file of known targets or motifs, such as those of 'Prosite' (Bairoch, 1990b) or user generated targets. Prosite is a dictionary of known patterns and motifs collected and extensively documented by Amos Bairoch. An example of a Prosite pattern is the consensus N-glycosylation site: N$[ST]$, where $ indicates any residue except P. Finding a Prosite in a new sequence provides a great deal of information to the researcher. It has been suggested that a search for Prosites in a new sequence will soon replace homology searches against the entire database as the first operation performed on a novel sequence. Once Prosite is available as a machine readable document, this planned implementation can procede. An obvious advantage is that a user could keep a file of currently interesting targets and then quickly search for all of them in any sequence of interest. An obstacle to such a development is the absence of an accepted standard syntax for targets to be discussed later.

A further development underway is the exploitation of connections between SwissProt and the corresponding DNA coding sequences. SwissProt, which is largely translated from DNA sequences, provides the possibility of simultaneous analysis of DNA and protein sequences. Within SCRUTINEER it will be possible to find out which DNA sequence corresponds to a particular protein sequence. Then it will be possible to address conveniently questions such as whether amino acids are coded differently near the N or C-terminii or where 'rare' codons occur and their relationship to domain boundaries (Krasheninnikov *et al.*, 1989). Another idea which is still in the planning stage is to provide convenient links to the Brookhaven database of three dimensional protein structures. Whenever a search is performed, it is of interest if the target occurs in a known tertiary structure and such a link will provide this information.

Pattern Exploration and Evaluation

Pattern searching is generally a cyclic process (Fig. 1). An initial target is used to perform a search. The initial target can have many sources but typical sources are aligned sequences or laboratory evidence suggesting that a particular region of a protein is important. The initial search results in hits. The user then screens the hits. It has been suggested that the screening might be automated (Lathrop *et al.*, 1987) but it is very difficult to encode in an algorithm all the knowledge of associations that a biologist has collected through years of experience. The process can then be terminated or the target can be modified based on the hits and a second search performed. The cycle of modification and searching may be carried out several times as the researcher learns about the salient features of the target and/or attempts to eliminate false positives. Once the target has been refined, the researcher may attempt to verify that it is 'good' in some sense. Some desirable qualities would be: (1) the target should be as simple and general as possible; (2) it should find few false positives and most of the known true positives; and (3) it should make sense biologically. The first and third points are particularly important. It is often possible with enough manipulation to make a

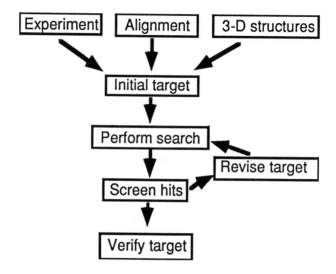

FIGURE 1. The cyclic process of pattern searching. Once the target has been sufficiently revised, an attempt is made to verify the target. The initial target can have many sources.

target specific. The worker should always check that the target has not become too 'contrived' or 'artificial'.

A common problem is deciding which of several targets to report when they all work equally well. In this case simplicity is a good guide. Another useful approach is to slightly alter each target and use the altered targets to perform searches. The most stable target will behave well even when altered. Since SCRUTINEER 'remembers' hits and can re-search those entries that contained hits, it is ideal for this type of work.

Examples of Motif Development

SCRUTINEER has been used recently by several groups and their work provides some interesting examples of exploration and evaluation. Kaupp *et al.* (1991), in an investigation of mononucleotide binding proteins, were able to describe a pattern:

[NDEQRK]G[DEA]X[AG]XXX[FY]XXXXG{15 35}GE{5 20}RXA

which located all cAMP/cGMP-dependent kinases in PIR release 22.0 with one false positive. The target syntax is: any one of the amino acids enclosed in square brackets is allowed at a given position, X is any amino acid and numbers in curly brackets indicate a sequence segment with an upper and lower bound on its length. The pattern is both simple and biologically reasonable in that the multiple amino acids allowed at a position are known to be similar biochemically.

Sometimes more complex patterns may emerge. Such an example, albeit speculative, is that of Gibson *et al.* (1989) working from an alignment of Rop and other helix-loop-helix proteins. Rop is small dimeric RNA-binding protein involved in regulating plasmid copy number. Here SCRUTINEER was used to try to determine if the alignment was feasible. To this end a pattern was constructed:

[RLMS][EA][RK]{0 1}[RKQI]@[EKQNARS]@@[NKRL]$$[FL]@@L$@@@{7 24}
[KE]@$@@@@@@@[YVFCH]@$$$.

In this case the {0 1} was deletion or a $. The variable length gap {7 24} corresponds to the putative loop region, @ means no P or G and $ means no G. The discrimination against P and G is reasonable since these are often avoided in helices. However there are several features of this pattern that make it appear contrived. Specifically, unlike the previous example, where multiple amino acids are allowed, they are quite dissimilar biochemically. Nonetheless the pattern, when used to search SwissProt 13.0, found no false positives and all true positives. When the three weakest positions [RLMS], [RKQI] and [NKRL] were changed to X then 26 false positives were found. This was deemed acceptable but this pattern is less certain that the first example.

The third example involved an attempt to relate all type I domains in the blue copper proteins (Ouzounis and Sander, unpublished). This family is quite divergent and the task therefore challenging. With SCRUTINEER one can specify that over a particular sequence segment each amino acid has a particular numerical value relative to some residue property such as hydrophobicity (see Nakai *et al.*, 1988 for an extensive collection of such properties), and a function score must lie within a specified range. This enables secondary structure predicted regions to be searched with the use of secondary structure preferences for the various amino acids (Fasman, 1989). The pattern that was developed required the addition of anti-parallel beta strand preference (Lifson and Sander, 1979) in order to be specific. With this addition though, the pattern was completely specific; there were no false positives and no true negatives. The anti beta strand preference is not contrived since the tertiary structural data indicate that it is appropriate. The resulting pattern is of high quality and may be expected to identify new blue copper proteins as they appear in the databases.

SCRUTINEER has proven useful in validating the uniqueness of search patterns and provides easy recognition of distant sequence family members in present and future database releases.

Current Issues

At the meeting which prompted this book discussion led to agreement on the need for a standard pattern syntax. Some consensus has already emerged primarily because programmers want to limit the amount users must learn to use their programs. For example, in SCRUTINEER, square brackets [], enclose residues that can all occur a single position. The same convention is also used by PSQ and GCG and Prosite.

As machine readable files of patterns become available, the need for a common syntax will become more pressing. The alternative will be similar to the currently ridiculous situation plaguing sequence databases for which there are numerous formats.

Availability of SCRUTINEER

SCRUTINEER has been documented and the program and the documentation are being distributed at no cost. The authors prefer to distribute source code via e-mail and let users compile it. The standard version is available on the EMBL file server (Stoehr and Omond, 1989) or from SIBBALD@EMBL.bitnet. Those lacking a Pascal compiler should inquire about the possibility of obtaining an executable image (for VAX computers), possibly on another medium. By the time this paper is published, the VAX version will also be available on the fileserver or directly from Sibbald.

Acknowledgements: We wish to thank Christos Ouzounis and Chris Sander, Drs. Kaupp, Boenigk, Luehring, Cook, Molday, Stuehmer , Martin Vingron, Toby Gibson and Peter Rice for making work available to us prior to publication. PRS is grateful for support from the Alexander von Humboldt-Stiftung and from the Natural Sciences and Engineering Research Council of Canada.

References

Altschul, S. F., Carroll, R. J., and Lipman, D. J. (1989). Weights for data related by a tree. *J. Mol. Biol.*, 207:647–653.

Bairoch, A. (1990a). *The SwissProt protein sequence databank user manual.* EMBL Data Library, Heidelberg, FRG. Release 14.

Bairoch, A. (1990b). *PROSITE: a dictionary of protein sites and patterns.* Univ. of Geneva. 5th release.

Bleasby, A. J. and Wootton, J. C. (1990). Construction of validated, non-redundant composite protein sequence databases. *Prot. Eng.*, 3:153–159.

Claverie, J.-M. and Bricault, L. (1986). *Prot. Struct. Funct. Genet.*, 1:60–70.

Devereux, J., Haeberli, P., and Smithies, O. (1984). A comprehensive set of sequence analysis programs for the VAX. *Nucl. Acids Res.*, 12:387–395.

Fasman, G. D. (1989). The development of the prediction of protein structure. In Fasman, G. D., editor, *The Prediction of Protein Structure and the Principles of Protein Conformation.* Plenum Press, N. Y. & London.

Gibson, T., Sibbald, P. R., and Rice, P. (1989). Rop/helix-loop-helix similarity. Unpublished results.

Gribskov, M., McLachlan, A. D., and Eisenberg, D. (1987). Profile analysis: detection of distantly related proteins. *Proc. Natl. Acad. Sci. USA*, 84:4355–4358.

Kaupp, U. B., Boenigk, W., Luehring, H., Cook, N. J., Molday, R. S., Stuehmer, W., and Vingron, M. (1991). The cyclic GMP-gated channel of rod photoreceptors. In press.

Krasheninnikov, I. A., Komar, A. A., and Adzhubei, I. A. (1989). The role of redundancy of the genetic code in determining the mode of cotranslational protein folding. *Biochemistry*, 54:187–200.

Lathrop, R. H., Webster, T. A., and Smith, T. F. (1987). Ariadne: pattern-directed inference and hierarchical abstraction in protein structure recognition. *Commun. ACM.*, 30:909–921.

Lifson, S. and Sander, C. (1979). Antiparallel β-strands differ in amino acid residue preference. *Nature*, 282:109–111.

Nakai, K., Kidera, A., and Kanehisa, M. (1988). *Protein Engineering*, 2:93–100.

Ouzounis, C. and Sander, C. (1990). A structure-derived sequence pattern for the type I domain in the blue copper proteins. Unpublished results.

Rice, P. (1990). PirOnly. Contact at EMBL, Meyerhofstr. 1, 6900 Heidelberg, FRG.

Sibbald, P. R. and Argos, P. (1990a). SCRUTINEER: a computer program which flexibly seeks and describes motifs and profiles in protein sequence databases. *CABIOS*, 6:279–288.

Sibbald, P. R. and Argos, P. (1990b). Weighting aligned protein or nucleic acid sequences to correct for unequal representation. Unpublished results.

Sidman, K. E., George, D. G., Barker, W. C., and Hunt, L. T. (1988). The Protein Identification Resource (PIR). *Nucl. Acids Res.*, 16:1869–1871.

Stoehr, P. J. and Omond, R. A. (1989). The EMBL network file server. *Nucl. Acids Res.*, 17:6763–6764.

Vingron, M. and Argos, P. (1989). A fast and sensitive multiple sequence alignment algorithm. *CABIOS*, 5:115–121.

Patterns and Specificity — Nucleotide Binding Proteins and Helicases

C. Hodgman

University of Cambridge
Department of Biochemistry
Tennis Court Rd.,
Cambridge CB2 1QW
U.K.

tch2@uk.ac.cambridge.phoenix

Conservation Patterns

In the folded protein structure hydrophobic residues tend to be buried and tightly packed while charged residues are excluded from the core. Consequently, hydrophobic residues appear to be better conserved than residues on the surface which do not have such severe constraints on their nature. Proteins also have ordered secondary structures: α-helices, β-strands (in sheets) and various loop structures (see Fig. 1). These different secondary structures cause consecutive residues in a sequence to be buried in different, but regular, ways. Work by Leszczynski and Rose (1986) has shown that active site residues tend to be in loops. For example, in the immunoglobulins, three loops come together to form a pocket whose specificity depends on the distribution of residues in these loops. Looking at the globins, Chothia and Lesk (1986) have shown that secondary structures shift with respect to each other in such a way that the active site or functional residues maintain the same three dimensional conformation, and that it is only in very divergent members of a family that some of these secondary structures begin to change. Because of packing constraints, insertions and deletions usually occur in the loops between the pieces of ordered secondary structure (that are not part of an active site).

The consequence of the above effects is that certain regions and residues are more conserved than others, resulting in a hierarchy of conservation. The active/functional site residues are most strongly conserved and are often the only residues that must be totally conserved. The residues that seem to be next in importance are those that are crucial for the correct folding and finally those involved in packing. Take the example

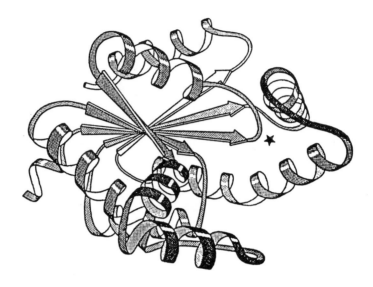

FIGURE 1. Schematic representation of the tertiary structure of adenylate kinase (Schultz *et al.*, 1986). The β-strands appear as arrows flanked above and below by α-helices. The phosphate binding loop is indicated by a star.

of the globins, there are two residues that are totally conserved: one is a histidine that is involved in haem binding (active site residue) and the other is a phenylalanine that is crucial for forming a bend at a particular point in the structure (key structural residue). There are also several positions where there are only a few residue types which correspond to regions important for packing, or in one case where proline is usually found that is probably important for forming a bend.

Structural Sources of Patterns

Broadly speaking, sequence patterns have been idintified by four means. Because patterns ought to relate to the polypeptide conformation, the best and most reliable sources are known tertiary structures. These have been determined mostly by X-ray crystallography but nuclear magnetic resonance and, recently, electron microscopy (with image processing) have also been used. Comparison of related structures or modelling homologous sequences into the known structure also provides information on the nature of the conservation at each sequence position (see Blundell *et al.*, in this Vol.). However, known structures are in short supply, so people have more often used

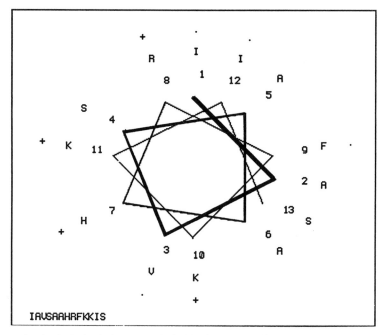

FIGURE 2. A helical wheel of the calmodulin-binding domain of myosin light-chain kinase. Note the left-hand face is positively charged compared with the right-hand face which is hydrophobic.

other approaches. Direct biochemical methods take a variety of forms which all aim to localise functional regions of the polypeptide. These methods include ligand binding, chemical crosslinking, and antibody binding/inactivation studies in conjunction with controlled proteolysis, and nowadays in vitro mutagenesis. From an alignment of the protein sequences in these key regions, patterns of amino acid distribution have been seen. The signal for N-glycosylation is probably the best known example, while others include the phosphorylation sites for the serine/threonine and tyrosine kinases, or the localisation signals for the nucleus or endoplasmic reticulum.

Some workers have carried out such biochemical experiments supplemented by computer analysis of the family to see what the sequences have in common. For example; the localisation signal for mitochondria has been interpreted as an α-helix (Schatz, 1987), as has the calmodulin-binding site of protein kinases (Fig. 2) for which there is support from spectroscopy and analysis of synthetic peptides (DeGrado *et al.*, 1987).

A similar form of analysis involves searching the alignment for positions where the residues have some common biophysical feature such as hydrophobicity or side chain volume. Examples of this approach include the cleavage site for signal peptidase (Folz

```
LVSNGTVFGIYRAN        IBV F2
::: :: :::::: :
LVSIGTIFGIYRKN        Fructose-1,6-bisphosphatase
```

FIGURE 3. An alignment of segment 1013–1024 of the IBV F2 open reading frame and residues 129–142 of porcine fructose-1,6-bisphosphatase.

and Gordon, 1987), sites for tyrosine sulphation (Hortin *et al.*, 1986), or 'acid blobs' — regions rich in acidic residues which play a role in gene activation. Such patterns tend to be quite vague and are mainly of value in identifying regions in a protein already known to possess a particular activity. They usually involve some special kind of computation, and to include them in a general database of sequence patterns would require the search algorithms to be included as well as the pattern itself.

The main source of patterns, however, is from the alignment of sequences determined by nucleotide sequencing. I will not discuss details of alignment techniques as they are the subject of other chapters in this book. However, a central problem with alignments is their credibility, especially when they become fragmentary, small, or vague. For example, when searching for putative functions of the major nonstructural protein of Infectious Bronchitis Virus (IBV), I found what seemed a rather interesting pairwise match with the AMP-binding region of Fructose-1,6-bisphosphatase. The probability of this match occurring by chance was an astonishing figure 2.4×10^{-14}, which appears fairly reliable. However, it was small and only two sequences were involved, so I examined the bisphosphatase literature and found that the lysine residue which mediates AMP binding is one of the few residues that is different in the IBV sequence. Consequently, the virus is most unlikely to have AMP-binding activity in this region and I am always cautious about putting too much faith in statistical values of any kind. I prefer to see some supportive biochemical data.

Nucleotide Binding Patterns

Patterns may be quite complex objects. They can consist of one or more short blocks of sequence defined in various ways which are then put together using the rules of Boolean logic (AND, OR, NOT). This section concerns how the individual blocks may be defined. This could be by the algorithms mentioned in the previous section

$$\{a_1..a_n\} \text{ AND } \{a_1..a_n\} \text{ AND } \left(\{a_1..a_n\} \text{ OR } \left(\{a_1..a_n\} \text{ AND } \{a_1..a_n\}\right)\right)$$

FIGURE 4. Structure of sequence patterns

(and elsewhere in this Vol.), but often the definition is simply derived from a sequence alignment.

Fig. 5 shows an alignment of subsequences common to many nucleotide triphosphate binding proteins, known variously as the Walker 'A' consensus, the GKS motif and the G-rich or P (for phosphate) loop. It will be used here to describe the methods of pattern definition.

The simplest way to define a motif is by considering only the positions where a residue is totally conserved. This is definition 'by consensus', but in this instance would include only the dipeptide GK. Such a definition would have little or no value in determining the function of an unknown sequence because this dipeptide is found in many more sequences than those involved in NTP-binding.

An alternative is to define by 'membership of set', which means that for each of a given set of positions only the residues found at that position are accepted. In our example, the boxed positions are often used to define the motif in this way. Other workers have modified this approach to good effect by categorising amino acids into groups and then when one member of a group is seen then every member of the group is permissible at that point (Taylor, 1986).

Alignment positions often show some preference for one or two amino acids which gets ignored if only membership-of-set is considered. To take this preference into account, various kinds of 'weight matrix' have been devised. This subtlety may still have shortcomings, however, especially when the number of sequences is limited. The observed distribution may not correspond to that of the 'motif family' in general.

Finally, recent attempts have been made to use neural networks to capture a pattern (Qian and Sejnowski, 1988), and other workers have defined patterns using tree methods normally used to make phylogenetic inferences (Smith and Smith, 1990). This method goes some way towards including information about groups of residues that tend to occur together in a given sequence — information which is often lost.

A large number of patterns have been found which relate in some way to known protein functions, and although their study may be interesting or illuminating, they will

ATP binding:

Protein	Residues	Sequence
E.coli ATPase α	164-182	RELII G DRGT GKT ALAIDA
E.coli ATPase β	144-162	KGVLF G GAGV GKT VNMMEL
malK protein	31-49	FVVFV G PSGC GKS TLLRMI
hisP protein	34-52	VISII G SSGS GKS TFLRCI
Myosin, rabbit	173-191	SILIT G ESGA GKT ENTKKV
Adenylate kinase	10-28	IIFVV G GPGS GKG TQCEKI

GTP binding:

G$_1$ α	35-53	KLLLL G AGES GKS TIVKQM
EF-Tu	13-31	NVGTI G HVDH GKT TLTAAI
EF-G	11-29	NIGIS A HIDA GKT TTTERI
p21 Ha-*ras*	5-23	KLVVV G AGGV GKS ALTIQL

Viral:

Rhino. 14 P2c	1224-1242	CVLIH G TPGS GKS LTTSIV
FMDV P2c	1212-1230	VVCLR G KSGQ GKS FLANVL
CPMV 58K	490-508	TIFFQ G KSRT GKS LLMSQV

FIGURE 5. An alignment of subsequences from some ATP and GTP binding and picornavirus proteins. The boxed positions denote those often used to define the motif. Adenylate kinase is slightly different in the second box and often discounted from sequence comparisons for that reason.

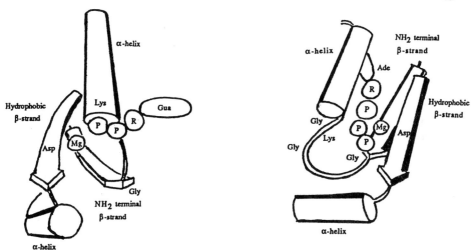

FIGURE 6. The nucleotide-binding subdomains of EF-Tu (or p21 Ha-*ras*) on the left and adenylate kinase on the right. The approximate positions of the nucleotide and magnesium ion are also shown.

remain putative until such time as we have structural data to back them up. Hence there is an urgent need to carry out some kind of biochemistry to support the hypotheses based on sequence patterns. So if any computer scientists reading this are not yet collaborating with bench scientists, I would recommend them to do so, because it is experiments which will ultimately confirm or deny any model you might produce.

Nucleotide Binding Pattern Specificity

There are many problems in defining motifs because the vaguer they become, the more members you are likely to find, but by the same token, more false positives will also be found. Because of my concern about using statistics as the sole measure of reliability, I recommend a more practical definition, which is simply the percentage of hits in a database search that turn out to be correct. A further indicator of a pattern's quality is how few false negatives there are — that is the number of sequences which should have been found but were not. However, the latter may be impossible to determine if very little is known about the pattern being studied.

To highlight these points, we can turn again to the GKS motif since we have structural information from Elongation Factor Tu (EF-Tu), Ha-*ras* (Jurnak, 1985; Pai *et al.*, 1989) and adenylate kinase (Fry *et al.*, 1985; Schultz *et al.*, 1986). The first two have the

Pattern scores

Pattern	Total entries	Entries (binding)	Entries (unk.)	Entries (nonbinding)
GxxxxGK(S,T)	248	113 (45.6%)	93 (37.5%)	42 (16.9%)
GxxxxGK(S,T) + hhhD	236	113 (47.9%)	92 (39.0%)	31 (13.1%)
weight matrix (from fig5)	142	84 (59.2%)	56 (39.4%)	2 (1.4%)

TABLE 1. Pattern searches were carried out on the PIR database (release 24) using SCRUTI-NEER (Sibbald and Argos, 1990, and also in this Vol.). The single-letter amino acid code is used, h denotes a hydrophobic residue (in this instance defined as A,I,L,M,V,F,W,Y,C,S, or T) while x refers to any amino acid. The weight matrix search used the default program settings on an alignment of the 8 residue segments bounded by the boxes in Fig. 5 (adenylate kinase was not included). Apart from the total number of matching entries, the numbers (and precentages in brackets) of known nucleotide binders, unknown (or putative) binders, and nonbinding proteins are also shown.

same overall fold — an encouraging sign that sequence pattern analysis can work since they were predicted to be so. The third has a very different overall fold and the G-rich loop is extended by a few residues (with the serine/threonine position substituted by a glycine), though the functional pieces are the same. From the N terminus of these protein domains, there is a β-strand which becomes the G-rich loop, and then there is an α-helix. At some later stage in the fold, there is a hydrophobic β-strand with and aspartate near its C-terminus. It is this aspartate (via a magnesium ion) and the totally conserved lysine which mediate binding to phosphates. It is worth noting firstly that this combination of secondary structures, called a Rossmann Fold (Rossmann et al., 1974), is found in other mono and dinucleotide binding proteins though there is no clear similarity in their primary sequences (for example aminoacyl tRNA synthetases and dehydrogenases); and secondly that these motifs merely identify the protein as triphosphate-binding while telling us nothing about which nucleotide base is bound, or what the overall function of the protein is.

Various people have cited motif I as an indicator of nucleoside triphosphate binding, whereas the above results clearly show that this is true in at best 80% of cases. The pattern is seen in the DNA polymerase of Epstein-Barr Virus and been quoted in the literature as a triphosphate site. However, the pattern occurs in a loop between two conserved regions and work on other herpesvirus DNA polymerases has shown that triphosphate binding takes place elsewhere in the structure (Larder et al., 1987). Clearly, great caution is required and it is important not to leap to any conclusions

```
EBV              YVGVLTDGKTL-MKGV
HSV              YIGVIYGGKML-IKGV
VZV              YIGVIYGGKVL-MKGV
CMV              YIGKVEGASGLSMKGV
Phage Phi29      YLRQKTYIQDIYMKEV
```

FIGURE 7. The EBV segment matching Motif I has been underlined.

when the only available data arise from sequence similarity.

Results from pattern matching methods offers guidelines for the molecular modeller. Successful examples include the work on Ha-*ras* mentioned bove (Halliday, 1983), the HIV proteinase (Pearl and Taylor, 1987; Blundell *et al.*, 1988), and antibody structures including the antigen binding loops (Chothia *et al.*, 1989). When the pattern of residue conservation in a herpesvirus family matched that of a protein kinase family, it prompted McGeoch and Davison to suggest that the herpes protein had the same function. Later experiments confirmed this (Frame *et al.*, 1987; Purves *et al.*, 1987).

Helicases

Pattern analysis also offers guidelines for determining the function of proteins by suggesting possible activities to test biochemically, and I will briefly give an example of this from my own work on helicases (Hodgman, 1988).

I was concerned with trying to find the functions and structure of the Tobacco Mosaic Virus replicative protein p126. General sequence comparisons showed that it was related to the equivalent proteins of other RNA viruses, and their alignment has several conserved regions including the two described in the previous section. When patterns corresponding to these conserved regions were used to search the protein database, they were found in the same order and sometimes with the same spacing in *E. coli* uvrD (HelicaseII). the biological significance of the similarity appeared dubious since the RNA viruses do not have a DNA intermediate. However, as other *E. coli* sequences related to uvrD were published, it became clear that the RNA virus and *E. coli* families share the same conserved regions. Moreover, a recently identified herpesvirus (thought to be DNA helicases) contained these regions as well, creating the impression of a superfamily.

Evans *et al.* (1985) have shown that the TMV p126 can crosslink azidoATP and I have preliminary data to show that p126 can be selectively eluted from cibacron

```
Motif          I                    Ia                          II                III                 IV            V                               VI      s

AlMV    821 VDGVAGCGKTNIK                             55 RLIFDECFLQH 15 VIGFGDTEQIPF 22 ITWRSPADA 66 IFTTHE-AGK-TFDNVYFCR 19 NGIVALSRH
BMV     687 VDGVAGCGKTTAIK                            54 RLLVDEAGLLH 15 VLAFGDTEQISF 22 KTYRCPQDV 78 IKTVHE-AGI-SVDNVTLVR 13 YCIVALTRH
CMV     709 VDGVAGCGKTTAIK                            54 RVLVDEVVLLH 15 ALCFGDSEQIAF 22 TTFRSPQDV 79 IKTVHE-SQGI-SEDHVTLVR 13 YCIVAVTRH
TMV     829 VDGVPGCGKTKEIL                            57 RLFIDEGLMLH 15 AXVVGDTQQIPY 24 TTLRCPADV 62 VHTVHE-VQGE-TYSDVSLVR 14 HVIVALSRH
ToMV    829 VDGVPGCGKTKEIL                            57 RLFIDEGLMLH 15 AXVVGDTQQIPY 24 TTLRCPADV 62 VHTVHE-VQGE-TYADVSLVR 14 HVIVSLSRH
TRV     901 VDGVPGCGKSTMIV                            56 VLHFDEALMAH 15 CICQGDQNQISF 24 ETYRSPADV 64 VSTVHE-SQGE-TFKDVVLVR 13 YLIVALSRH
SFV     183 VFGVPGSGKSAIIK                            50 ILYVDEAFACH 16 VVLCGDPKQCGF 21 ISRRCTRPV 58 VMTAAA-SQGI-TRKGVYAVR 14 HVNVLLTRT
SV      183 VIGTPGSGKSAIIK                            50 VLYVDEAFACH 16 VVLCGDPMQCGF 24 ISRRCTQPV 58 VMTAAA-SQGI-TRKGVYAVR 14 HVNVLLTRT

IBV    1209 VQGPPGSGKSHFAI                            54 ILLVDEVSMLT 15 VVYVGDPAQLPA 30 KCYRCPKEI 82 VQTVDS-SQGS-EYDYVIFCV 11 RFNVALTRA

BNYVV1  893 VKGGPGTGKSFLIR                            48 IIFVDEFTAYD 11 IYLVGDEQTGI 25 MNFRNPVHD 72 KTTVRA-NQGS-TYDNVLPV 12 LNIVALSRH
BNYVV2  121 VLGAPGVGKSTSIK                            49 TMLVDEVTRVH 11 VICFGDPAQGLN 18 ASRRFGKAT 67 SILYSD-AHGQ-TYDVVTIII 13 VRAVLLTRA
BSMV2   267 ISGVPGSGKSTIVR                            41 LLIIDEYTLAE 11 VLLVGDVAQGKA 18 TTYRLGQET 62 CALAID-VQGK-EFDSVTLFL 12 LRIVALSRH

uvrD     26 VLAGAGSGKTRVLV 17 MAVTF-TNKAAAEMRHRI 140 NILVDEFQNTN 16 VMIVGDDDQSIY 26 QNYRSTSNI 267 LMTLHS-AKGI-EFPQVFIVG 23 LAYVGVTRA
rep      19 VLAGAGSGKTRVIT 17 AAVTF-TNKAAREMKERV 141 YLLVDEYQDTN 16 FTVVGDDDQSIY 26 QNYRSSGRI 271 LMTLHA-SKGI-EFPYVVMVG 22 LAYVGLTRA
recB     20 IEASAGTGKTFTIA 25 LVVTF-TEAATAELRGRI 303 VAMIDEFQDTD 18 LLLIGDPKQAIY 24 TNWRSAPGM 286 IVTIHK-SKGI-EYPLWWIPF 44 LLYVALTRS
recD    164 ISGGPGTGKTTTVA 17 RLAAP-TGKAAARLTESL  48 VLVVDEASMID 16 VIFLGDRDQLAS 24 QLSRLTGTH 198 AMTVHK-SQGS-EFDHAALIL 11 LVTAVTRA

EBV      69 ITGTAGAGKSTSVS 7 CVITGTTVIAAQNLSAIL  88 VIVVDEAGTLS 26 IVCVGSPTQTDA 44 NNKRCTDVQ 426 AMTIAK-AGGI-SLNKVAICF 9 HYVVALSRA
HCMV    117 VTGTAGAGKTSSIQ 7 CLVTGATTVAAQNLSQTL 100 IIVIDECGLML 26 IICVGSPTQTEA 44 HNKRCTDLD 513 AMTIAK-SQGI-SLEKVAVDF 10 HIYVAMSRV
HSV      94 ITGNAGSGKSTCVQ 7 CVVTGATRIAAQNMYAKL 111 VIVIDEAGLLG 26 LVCVGSPTQTAS 44 NNKRCVEHE 442 AMTITR-SQGI-SLDKVAICF 8 SAYVAMSRT
VZV      87 ISGNAGSGKSTCIQ 7 CIITGSTRVAAQNVHAKL 110 VIVIDERAGLLG 26 IVCVGSPTQTDS 44 NNKRCQEDD 447 AMTIAR-SQGI-SLEKVAICF 8 SVYVAMSRT

PIF     255 YTGSAGTGKSILLR 7 VAVTASTGLAACNIGGTI  21 ALVVDEISMLD 25 LIFCGDFFQLPP 29 KVFRQRGDV 219 MQTIHQNSAGKRRLPLVRFKA 33 QAYVALSRA

Residue      V G AG GKS    * * **   VT  T  AA N   L       VDE    ***** *     GD Q      R       *** *** E  V     *******
distribn.    I A P    T       IA   T E      I   I          VDE      I           GD Q      R         AK S  T  A        VALSR
             Y          A      R      V                     F            L                           L    VH   R               TLVT
                                                                                                          NA                     GM
                                                                                                                                 SI
```

FIGURE 8. The alignment of conserved regions from known and putative helicases as described in Hodgman (1988).

blue columns by ATP and UTP. This gives some credibility to the notion that the RNA virus family are helicases. There are a total of 6 conserved segments, though the sequences implicated in DNA unwinding share a seventh (called Motif Ia). It is the large difference in size of the region between Motifs I and II that make me wary of ascribing RNA helicase activity to the virus domain. However, since I finished that work, a separate RNA helicase family (see Table 2) has come to light which clearly shares some of the motifs already described. In particular, Motifs I, Ia, II, V and VI are readily recognisable while motifs III and IV are poorly defined if present at all.

These observations show that helicases play a role in many cellular functions and many have common structural elements if not a common fold. However, not all helicases fit into the sequence patterns described, which suggests that helicase activity may be achieved by more than one kind of protein fold. Conversely, there are many otherwise independent sequence families which overlap the helicase families in terms of some of the motifs they share.

Conclusions

I have found some confusion in the litarature over the precise definition of various words (pattern/motif/template/fingerprint). It would be helpful to have a common terminology so that we know what we mean when communicating with one another. Similarly, it would be preferable if we could choose standard definitions for patterns, although that would be much more problematic. Several groups have already started organising and distributing collections of patterns, and in this it will be important to have some kind of quality control. Whilst some patterns might look promising or useful, more careful examination often reveals them to be less specific than expected.

Perhaps most importantly, we need to communicate with the people outside the computer-based community to sort out access and distribution of databases and software to the bench scientist in particular — who for the large part does not understand where you can and cannot put trust in all these things. Finally, as I have demonstrated, we need to work on some way of presenting pattern data that is neither a sea of sequence characters nor something too simplistic.

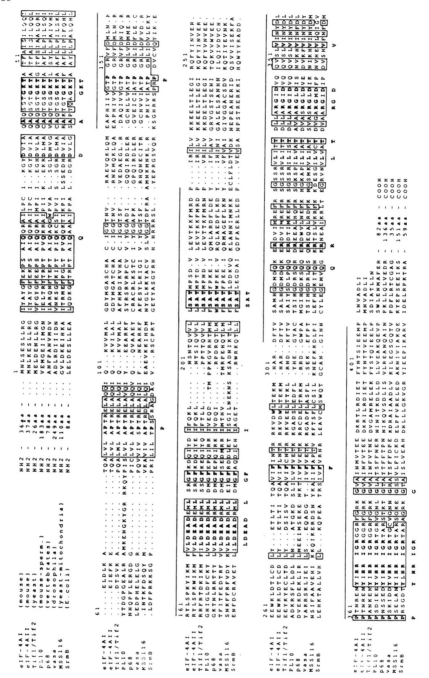

TABLE 2. 'DEAD box' sequences. This list is taken from Linder *et al.* (1989).

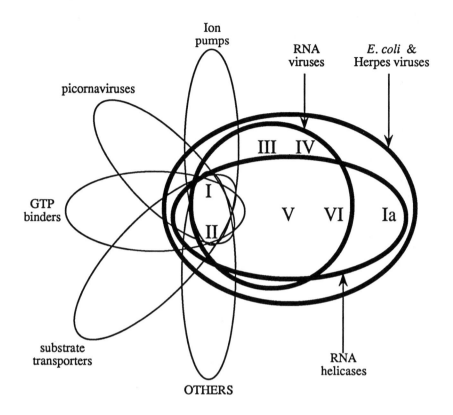

FIGURE 9. The seven motifs fall into overlapping sets (denoted by thick lines) belonging to the three families. However, Motifs I and II (involved in phosphate binding) are also common to several other families which have other conserved regions of their own.

References

Blundell, T., Carney, D., Gardner, S., Hayes, F., Howlin, B., Hubbard, T., Overington, J., Singh, D. A., Sibanda, B. L., and Sutcliffe, M. (1988). Knowledge-based modelling and design. *Eur. J. Biochem.*, 172:513–520.

Chothia, C. and Lesk, A. M. (1986). The relation between divergence of sequence and structure in proteins. *EMBO*, 5:823–826.

Chothia, C., Lesk, A. M., Tramontano, A., Levitt, M., Smith-Gill, S. J., Air, G., Sheriff, S., Padlan, E. A., Davies, D., Tulip, W. R., Colman, P. M., Spinelli, S., Alzari, P. M., and Poljak, R. J. (1989). Conformations of immunoglobulin hypervariable regions. *Nature*, 342:877–883.

DeGrado, W. F., Erikson-Viitanen, S., Wolfe, Jr, H. R., and O'Neil, K. T. (1987). Predicted calmodulin-binding sequence in the gamma subunit of phosphorylase b kinase. *Proteins*, 2:20–33.

Evans, R. K., Haley, B. E., and Roth, D. A. (1985). Photoaffinity labeling of a viral induced protein from tobacco. *J. Biol. Chem.*, 260:7800–7804.

Folz, R. J. and Gordon, J. I. (1987). Computer-assisted predictions of signal peptide processing sites. *Biochem. Biophys. Res. Commun.*, 146:870–877.

Frame, M. C., Purves, F. C., McGeoch, D. J., Marsden, H. S., and Leader, D. P. (1987). Identification of the *Herpes simplex* virus protein kinase as the product of viral gene US3. *J. Gen. Virol.*, 68:2699–2704.

Fry, D. C., Kuby, S. A., and Mildvan, A. S. (1985). NMR studies of the MgATP binding site of adenylate kinase and of a 45-residue peptide fragment. *Biochemistry*, 24:4680–4694.

Halliday, K. R. (1983). Regional homology in GTP-binding proto-oncogene products and elongation factors. *J. Cyclic Nucleotide Protein Phosphorylation Res.*, 9:435–448.

Hodgman, T. C. (1988). A superfamily of replicative proteins. *Nature*, 333(22,23):578.

Hortin, G., Folz, R., Gordon, J. I., and Strauss, A. W. (1986). Characterisation of sites of tyrosine sulfation and criteria for predicting their occurrence. *Biochem. Biophys. res. Commun.*, 141:326–333.

Jurnak, F. (1985). Structure of the GDP domain of EF-Tu and location of the amino acids homologous to *ras* oncogene proteins. *Science*, 230:32–36.

Larder, B. A., Kemp, S. D., and Darby, G. (1987). Related functional domains in virus DNA polymerases. *EMBO*, 6:169–175.

Leszczynski, J. F. and Rose, G. D. (1986). Loops in globular proteins: a novel category of secondary structure. *Science*, 234:849–855.

Linder, P., Lasko, P. F., Ashburner, M., Leroy, P., Nielsen, P. J., Nishi, K., Schnier, J., and Slominski, P. P. (1989). Birth of the D-E-A-D box. *Nature*, 337:121–122.

Pai, E. F., Kabsch, W., Krengel, U., Holmes, K. C., John, J., and Wittinghofer, A. (1989). Structure of the guanine-nucleotide-binding domain of Ha-*ras* oncogene product p21 in the triphosphate conformation. *Nature*, 341:209–214.

Pearl, L. H. and Taylor, W. R. (1987). A structural model for the retroviral proteases. *Nature*, 329:351–354.

Purves, F. C., Longnecker, J. M., Leader, D. P., and Roizman, B. (1987). *Herpes simplex* virus 1 protein kinase is encoded by open reading frame us3 which is not essential for virus growth in cell culture. *J. Virol.*, 61:2896–2901.

Qian, N. and Sejnowski, T. J. (1988). Predicting the secondary structure of globular proteins using neural network models. *J. Mol. Biol.*, 202:865–884.

Rossman, M. G., Moras, D., and Olsen, K. W. (1974). Chemical and biological evolution of nucleotide binding proteins. *Nature*, 250:194–199.

Schatz, G. (1987). Signals guiding proteins to their correct locations in mitochondria. *Eur. J. Biochem.*, 165:1–6.

Schultz, G. E., Schiltz, E., Tomasselli, A. G., Frank, R., Brune, M., Wittinghofer, A., and Schirmer, R. H. (1986). *Eur. J. Biochem*, 161:127.

Smith, R. F. and Smith, T. F. (1990). Automatic generation of primary sequence patterns from sets of related protein sequences. *Proc. Natnl. Acad. Sci. USA*, 87.

Taylor, W. R. (1986). The classification of amino acid conservation. *J. Theor. Biol.*, 119:205–218.

Discussion

C. Sander: I would like point out there is complete literature on sensitivity and specificity and we should not invent new statistical terms. The pattern and motif literature is now documented with proper statistical terms directly from epidemiology using sensitivity and specificity, it is just not referenced and used by people. As regards terminology, it is my experience that it sorts itself out.

T. Smith: Computer scientists have for many years used the term 'regular expression' which applies to the patterns we have been considering here.

R. Abarbanel: However, weight-matrix and neural-net patterns are not regular expressions. Hence the term does not apply to all kinds of patterns.

Q: On the basis of pattern analysis, are you confident your six-motif patterns always define helicases?

A: Well, all I've ever been able to do is say that the motifs are there. It could be a helicase signature, but I don't know. In the example of the viral sequences that fit the Walker concensuses, because they are for the most part conserved, I think it's probable that they do have triphosphatase activity but I'm awaiting experimental confirmation of that. I think when, in a database search you pull out one sequence, you then must see if there are any other sequences that are definitely related to it to see if the pattern you've found is conserved in the whole family. It was on that basis that I decided there is some relationship between these RNA virses and the *E. coli* proteins.

P. Argos: Do you think it is appropriate to grade patterns solely in terms of percentage specificity?

A: Well, I've only been using it as a practical measure. I've always calculated some statistics on top of that, but I find the statistical values that come out tend to be huge. And from experience I know to take them with a pinch of salt. The problem is that protein sequences are far from random. There are preferences for certain residues to come after others and so I've always preferred working with real sequences, rather than be one step removed and rely on statistics.

The Helix-Turn-Helix Motif and the Cro Repressor

Wayne F. Anderson

Department of Biochemistry
Vanderbilt University
School of medicine
Nashville, TN 37232
U.S.A.

anderson@vuhhcl01.edu

Introduction

Bacteriophage λ codes for two repressor proteins. They are the products of the cI gene which makes the λ Repressor and the product of Cro gene which makes the Cro Repressor. These proteins are part of a molecular switch between the two developmental pathways that are open to the phage when it infects an *E. coli* cell. Both of these repressor proteins function by binding at the left and right operator regions, OL and OR. These operator regions each have three operator sequences that are imperfect inverted repeats or palindromes and they are operationally defined as 17 base pairs in length. These two proteins bind to all six of the operator sequences, however, they bind with different relative affinity. Cro Repressor binds most tightly to OR3, and in doing so shuts off transcription from the promoter that is responsible for maintenance levels of the λ Repressor. On the other hand, the λ Repressor binds co-operatively to OR1 and OR2, and in binding to these operators it shuts off transcription from promoter PR.

Comparative Structural Studies

Some time ago, in collaboration with Brian Matthews and Yoshinori Takeda (Anderson *et al.*, 1981) we determined the structure of λ Cro protein (see Fig. 1). It is a small dimeric protein. Each subunit of the dimer is just 66 amino acid residues (Hsiang *et al.*, 1977). It is quite a simple protein comprised of three α-helices and a three-stranded anti-parallel β-sheet. An examination of the structure suggested that it might bind to DNA such that one helix of each subunit would be positioned in successive major grooves. The nice complementarity between an α-helix and the major groove of B-DNA had been recognized a long time ago (Zubay and Doty, 1959). Initially it was proposed that this was the way histones bound to DNA (Zubay and Doty, 1959; Sung and Dixon, 1970).

When the structure of the Cro Repressors was compared with that of the bacterio-phage λ Repressor (Pabo and Lewis, 1982; Ohlendorf et al., 1983), and with the *E. coli* CAP gene activator (McKay and Steitz, 1981; Steitz et al., 1982) we found that they all contained a particular structural motif, a helix, a characteristic turn and a second helix, in their proposed binding regions. Given the structural similarity in the motif it was suprising to find that in these dimeric proteins the motif was arranged in very different relative orientation with respect to the presumed location of the DNA. The region of similarity is quite short (20 to 24 residues) and α-helices are very common secondary structures, so a natural question is: is the similarity in the motifs significant or just coincidence?

The root mean square discrepancy in the 24 α-carbon position of the helix-turn-helix motif of Cro and that in CAP was found to be 1.1Å (Steitz et al., 1982), while that of the 23 residues segments of Cro and λ Repressor was 0.7Å (Ohlendorf et al., 1983). In contrast, when the 24 residues helix-turn-helix motif of Cro was compared with the proteins that were available in the Brookhaven Protein Data Bank (Bernstein et al., 1977) in 1982 the next best had an RMS discrepancy of 2.8Å (Steitz et al., 1982). Given that the coordinates were not refined at that time, the differences between the helix-turn-helix motifs in these three proteins approached the approximate errors in the coordinates. However, when both of the motifs in the Cro dimer were compared with both of the motifs in the λ Repressor dimer the RMS discrepancy rose to 3.4Å (Ohlendorf et al., 1983), pointing out the differences in the way the subunits are associated and the different relative orientations of the helix-turn-helix motifs. This was a surprise since Cro and λ Repressor bind the same DNA sequences — the six λ operators.

These comparisons led to a general model in which DNA binding proteins had helix-turn-helix motifs as part of a protein structure that could otherwise be very different. The helix-turn-helix units were expected to bind to the DNA with the second helix associated with the major groove and the B-DNA. The structural similarities then led us to look at the amino acid sequences (in fact we had been doing that beforehand but did not believe the results until we saw the structures and it was clear that the sequence comparison results were meaningful).

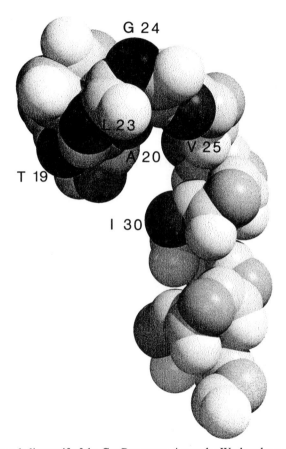

FIGURE 1. The helix-turn-helix motif of the Cro Repressor. A van der Waals sphere representation of the main chain atoms of residues 16–35 is shown. The first helix of the motif is coming toward the viewer, on the left, while the second, DNA recognition helix, is vertical with the side that interacts with the DNA on the far right. The α-carbons are slightly darker than the other atoms. The darkly shaded atoms are the α-carbons of Thr-19, Ala-20, Leu-23, Gly-24, Val-25 and Ile-30 (motif residues 5,6,9,10,11, and 16, see Fig. 2).

```
                        6         10            16
Cro        15 G Q T K T A K D L G V Y Q S A I N K A I H
Rep        32 S Q E S V A D K M G M G Q S G V G A L F N
434Cro     17 T Q T E L A T K A G V K Q Q S I Q L I E A
434Rep     17 N Q A E L A Q K V G T T Q Q S I E Q L E N
cII        25 G T E K T A E A V G V D K S Q I S R W K R
lac Rep     5 T L Y D V A E Y A G V S Y Q T V S R V V N
CAP       168 T R Q E I G Q I V G C S R E T V G R I L K
trp Rep    67 S Q R E L K N E L G A G I A T I T R G S N
Mat a1    115 E K E E V A K K C G I T P L Q V R V W V C
Lex A      27 T R A E I A Q R L G F R S P N A A E E H L
```

FIGURE 2. The amino acid sequences of the helix-turn-helix motif of a selected group of proteins.

Sequence Analysis and Motif Matching

The amino acid sequences of the helix-turn-helix regions of a small number of proteins are shown in Fig. 2. From an alignment like this we could predict where these helix-turn-helix motifs are in these other proteins. One of the reasons we initially were not sure about the alignment was that if you compared the amino acid sequence of Cro with the amino sequence of λ Repressor the distribution of scores was what you would expect from random sequences. Statistically there was no significant similarity between Cro and λ Repressor. However, as the numbers of sequences increased more cases such as the 434 Cro and 434 Repressor, which are very similar to each other were found. Also, the 434 Cro and Repressor proteins are very similar throughout the lengths of their sequences.

The beautiful structures from Harrison's lab of 434 Cro (Mondragon *et al.*, 1989b), the amino terminal, DNA binding fragment of the 434 Repressor (Mondragon *et al.*, 1989a) and the complexes of both of them with DNA (Wohlberger *et al.*, 1988; Aggarwal *et al.*, 1988) has verified that the region indicated on Fig. 2 is the helix-turn-helix motif and it is involved in binding to the DNA. Similarly, with the Trp Repressor, the structural studies from Sigler's lab show that the helix-turn-helix motif was correctly predicted (Schevitz *et al.*, 1985; Otwinowski *et al.*, 1988).

Protein tertiary structure is conserved more strongly than the primary sequence of the amino acids. Consequently, when you compare sequences, one question is where to assign a cutoff on their similarity. When faced with an essentially continuous distribution of scores, it is necessary to decide whether you want to miss sequences that should be included or to include sequences that should not be there. To do this in an objective way requires careful statistical analysis.

These sequence motifs were first found by pair-wise comparisons of the sequences and discovering that they could fit in a common alignment. The desire to locate structural similarities like the helix-turn-helix motif led us to develop a technique (Bacon and Anderson, 1986; Bacon and Anderson, 1990) derived from the fixed length sequence comparisons of Cantor and Jukes (Cantor and Jukes, 1966) or Fitch (Fitch, 1966) that would allow the detection of weak similarities among the number of sequences. This simple approach takes a segment from the first sequence and compares it with all possible segments of the second. This is repeated with successive segments until all possible segments have been compared. For each comparison, a score is calculated for the alignment by adding up scores reflecting individual amino acid similarities at each position in the segment. For pair-wise comparison this method requires computational time that is roughly proportional to the product of the sequence lengths. The generalization of this to a larger number of sequences can be expensive. If you now want to ask what is common among N sequences the amount of time that is spent would go up more or less like the average sequence length to the nth power. This quickly becomes impossible. One way to get around the problem is to only look at what are potentially the good comparisons instead of all of them. In doing this you must, however, make the assumption that if there is going to be anything in common that is statistically significant in the similarity of these sequences when the whole set is considered together, then, even if the similarity is not significant in a particular pair, it will still be among the better segment comparisons. If this is true, it is not necessary to keep all possible alignments of segments but only a list of the better ones and when a new sequence is added it is compared against the previous list of possible segment alignments. Tests of the method using random sequences indicated that good comparisons were unlikely to be missed. Further tests also have shown that adding sequences that do not have features in common with others in the group reduce the ratio of the observed to calculated frequency for the overall score (Bacon and Anderson, 1986).

The results for five DNA binding proteins, bacteriophage λ Cro, 434 Cro, λ Repressor, CII protein and the *E. coli* CAP gene activator protein are given in Table 1. The value given is a ratio of the observed frequency of the highest score divided by the expected frequency of that score for random sequences. There are a few pair-wise alignments that are quite good, such as λ Cro and 434 Cro or λ Cro and λ CII. But other pairs are no better than random sequences. However, when you ask what is

Proteins Compared	Ratio
Cro - Rep	7.7
Cro - CAP	1.0
Cro - 434	797.
Cro - cII	840.
Rep - CAP	2.8
Rep - 434	6.6
Rep - cII	3.1
CAP - 434	1.0
CAP - cII	2.5
434 - cII	0.9
Cro - Rep - CAP	11.
Cro - Rep - 434	210,000
Cro - Rep - cII	7,300
Cro - CAP - 434	182
Cro - CAP - cII	138
Cro - 434 - cII	15,000
Rep - CAP - 434	390
Rep - CAP - cII	31
Rep - 434 - cII	86
CAP - 434 - cII	4.6
Cro - Rep - CAP - 434	91,000
Cro - Rep - CAP - cII	8,300
Cro - Rep - 434 - cII	2,700,000
Cro - CAP - 434 - cII	14,000
Rep - CAP - 434 - cII	2,700
Cro - Rep - CAP - 434 - cII	3,600,000

TABLE 1. Ratio of observed to expected frequency for the best score on the comparison of DNA binding proteins.

common among all five of them together, the ratio of the frequency of the top score divided by the expected frequency goes up enormously. This indicates that there is a common region of sequence among them which is not obvious just from looking at the pairs. Using the observed distribution to derive probabilities is clearly impossible for this approach because there is not a complete distribution of scores. Only part of the distribution has been examined. To avoid this problem we calculate what the distribution should be using McLachlan's (McLachlan, 1971) double matching probability generalized to a larger number of sequences. In testing random and unrelated sequences, the ratio of observed frequency to expected frequency is very close to 1.0. When sequences are added that share some similarity, the ratio goes up.

Predicted Motifs

The sequences of the predicted helix-turn-helix motifs from a number of proteins are aligned in Fig. 2. For many of these proteins the structure is now known and the motif is in the predicted location. The multiple sequence comparison results can be used to look for patterns of residues. One thing that is common in all of the sequences shown in Fig. 2 is a glycine in the turn between the two helices. This residue adopts a conformation close to that of a left handed α-helix, a conformation which is less favorable if the resiude has a side chain. Hence glycine is favored at that position. However, it does not have to be a glycine and there are mutants of the λ Repressor which are functional and which have a glutamate at that position instead of a glycine.

Other features of the pattern include hydrophobic residues before and after the glycine. Also, the first helix of the motif generally has an alanine or a glycine at position 6 but again this is not essential and in the Trp Repressor that residue is a lysine. The lysine side chain has enough conformational flexibility that it can fit into the structure while the amino group still can get out to the solvent. A β-branched side chain is forbidden at this position. Again the residue just before the alanine is generally a hydrophobic residue. An exception is λ Cro where it is a threonine which is buried. The threonine can satisfy its hydrogen bonding potential by making a hydrogen bond to the backbone in the helix one turn back.

The second helix has a hydrophobic residue, generally isoleucine or valine at position 16. But the residues involved in contacting the DNA are mostly hydrophilic residues, and since these contact the sugar phosphate backbone they generally include a number of basic residues. When the Cro and λ Repressor structures are examined, the side chains in the motif that tend to be conserved (Fig. 1) are tightly packed in the core while the residues on the outer surfaces of the helices are quite variable.

The structure of the motif places some restraints on the amino acid sequences but it does not constrain it very much. There are no residues that are absolutely conserved but overall they are restrained by the structural pattern. Again, if the λ Repressor and Cro are examined the overall structures are quite different in the way that the recognition helix is oriented relative to its twofold symmetry axis (Ohlendorf *et al.*, 1983).

New Structural Studies

Within the last year or two there have been a number of structures of complexes of repressors with synthetic oligonucleotides. Two that are important to look at when thinking about going from sequence patterns to modelled structures are the 434 Cro (Wohlberger *et al.*, 1988) and 434 Repressor (Aggarwal *et al.*, 1988). These proteins have quite similar amino acid sequences, even outside the helix-turn-helix motif, bind the same operators and were crystallized with the same oligonucleotide.

When only one subunit of the 434 Cro was superposed on one subunit of the 434 Repressor the monomers were found to have very similar structures (RMS = 0.77Å). However, because of subtle differences in the two structures, the other subunits of these dimers end up not being superimposed very well at all (RMS difference of 3.3Å) (Wohlberger *et al.*, 1988). Consequently, these proteins which are very similar throughout their sequence, contact the DNA in different ways. The motif is conserved and yet the interaction with the DNA is different.

The implication of the results on 434 Cro and 434 Repressor is that if you can predict where the helix-turn-helix motif is and you know that it is involved in DNA binding, or even if you can predict the whole structure of the subunit, you still can not predict how the motif is going to interact with the DNA. Consequently, at the present time you cannot predict what are the interactions that give rise to the base sequence specificity. This implies that there is not a simple recognition code for how the protein recognizes the DNA.

Comparison of the 434 Cro and Repressor structure revealed small, but very significant differences in the way they recognize the same DNA sequence. The picture gets more confused when the Trp and Lac Repressors are considered. Even larger differences occur in the way the Trp Repressor is bound to the DNA (Otwinowski *et al.*, 1988). Once again the helix-turn-helix motif is very similar to what it is in Cro. However, in this case, the recognition helix is bound almost end on into the major groove of the DNA, rather than lying in the major groove of the DNA (Otwinowski *et al.*, 1988).

NMR studies of the Lac Repressor head piece (Kaptein *et al.*, 1985; Lamerichs *et al.*, 1989a) reveal that the conserved helix-turn-helix motif occurs as predicted. But it binds in an orientation 180 degrees away from the orientation that it has in Cro. So, again, you have the same motif occurring but it contacts the DNA in a totally different way.

The DNA binding domain of the lexA protein (Lamerichs *et al.*, 1989b) has a structure that at first appears similar to the helix-turn-helix motifs that have been discussed. However, when it was compared with the helix-turn-helix motif in the Lac Repressor, which is similar to the motif in other repressors, it was found that for lexA the second helix, the recognition helix, was oriented in a different manner relative to the first helix. It appears that the reason for this difference is that there is a phenyalanine next to the glycine in the turn and the turn has gained two residues from the second helix. Because the phenyalanine is larger than the equivalent residue in the other proteins and is positioned between the helices it forces them apart. Again this raises the question of how dissimilar two structures can be and still count as being a helix-turn-helix motif? As the sequences and structures get less and less similar how do you decide where to draw the line? Should it depend on biological function, RMS discrepancy in coordinates, or how well the particular sequence pattern fits?

A more recent structure comparison by Brennan and Matthews (Brennan and Matthews, 1989) of the helix-turn-helix motif and the contents of the Brookhaven Protein Data Bank showed that RMS differences between the observed helix-turn-helix motifs are quite small. On the other hand, there are now also fragments from a few other proteins that are structurally similar to the motif. Some of these were also recognized by Jane and David Richardson in a search of the protein databank and include the ribosomal protein L7/L12. Do these proteins belong in the same category? Many of the sequences that have this motif, particularly the phage proteins, have clearly diverged from some common evolutionary precursor as shown by the similarity in the DNA sequences of the genes. On the other hand, are these other proteins instances where the structures have converged because it is a convenient way to fold or because the α-helix is nicely complementary to the major groove of B-DNA?

As mentioned above, it is not surprising that α-helices commonly mediate recognition of B-DNA. With the helix-turn-helix proteins, it would seem that this motif is primarily a good way to hold the recognition surface of the protein in the right position and in the right orientation for interacting with the DNA. What seems unusual, especially when they are compared with enzymes, that tend to conserve overall folding patterns, is that the rest of the structure really determines how the motifs are positioned relative to each other and consequently determines how that motif contacts the DNA. Evolutionarily speaking, perhaps the simplest way to get different specificities was to alter the orientation of the motif with respect to the DNA.

The variation in the structures of these DNA binding proteins would make it very difficult to predict what interaction a protein is going to make with the DNA. There is no simple code. On the other hand, most of the proteins have incorporated some flexibility in the structure and there is also flexibility in the structure of the DNA. As Harrison's group found with the 434 Cro and 434 Repressor complexes it is possible to use the same oligonculeotide in the complexes and the DNA adopts different conformations depending on whether it is bound to the Cro protein or to the Repressor.

References

Aggarwal, A., Rodgers, D. W., Drottar, M., Ptashne, M., and Harrison, S. C. (1988). *Science*, 242:900–907.

Anderson, W. F., Ohlendorf, D. H., Takenda, Y., and Matthews, B. W. (1981). *Nature*, 290:754–758.

Bacon, D. J. and Anderson, W. F. (1986). *J. Mol. Biol.*, 191:153–161.

Bacon, D. J. and Anderson, W. F. (1990). *Meth. Enz.*, 183:438–447.

Bernstein, F. C., Koetzle, T. F., Williams, G. J. B., Meyer, E. F., Brice, M. D., Rogers, J. R., Kennard, O., Shimanouchi, T., and Tasumi, M. (1977). *J. Mol. Biol.*, 112:535–542.

Brennan, R. G. and Matthews, B. W. (1989). *J. Biol. Chem.*, 264:1903–1906.

Cantor, C. R. and Jukes, T. H. (1966). *Proc. Natl. Acad. Sci. USA*, 56:177–184.

Fitch, W. M. (1966). *J. Mol. Biol.*, 16:9–16.

Hsiang, M. W., Cole, R. D., Takeda, Y., and Echols, E. (1977). *Nature*, 270:275–277.

Jordan, S. R. and Pabo, C. O. (1988). *Science*, 242:893–899.

Kaptein, R., Zuiderweg, E. R. P., Scheek, R. M., Boelens, R., and van Gunsteren, W. F. (1985). *J. Mol. Biol.*, 182:179–182.

Lamerichs, R. M. J. N., Boelens, R., van der Marel, G. A., van Boom, J. H., Kaptein, R. J. H., Buck, F., Fera, B., and Rüterjans, H. (1989a). *Biochemistry*, 28:2985–2991.

Lamerichs, R. M. J. N., Padilla, A., Boelens, R., Kaptien, R., Ottleben, G., Rüterjans, H., Granger-Schnarr, M., Oertel, P., and Schnarr, M. (1989b). *Proc. Natl. Acad. Sci. USA*, 86:6863–6867.

McKay, D. B. and Steitz, T. A. (1981). *Nature*, 290:744–749.

McLachlan, A. D. (1971). *J. Mol. Biol.*, 61:409–424.

Mondragon, A., Subbiah, S., Almo, S. C., Drottar, M., and Harrison, S. C. (1989a). *J. Mol. Biol.*, 205:189–200.

Mondragon, A., Wohlberger, C., and Harrison, S. C. (1989b). *J. Mol. Biol.*, 205:179–188.

Ohlendorf, D. H., Anderson, W. F., Lewis, M., Pabo, C. O., and Matthews, B. W. (1983). *J. Mol. Biol.*, 169:757–769.

Otwinowski, Z., Schevitz, R. W., Zhang, R.-G., Lawson, C. L., Joachimiak, A., Marmorstein, R. Q., Luisi, B. F., and Sigler, P. B. (1988). *Nature*, 335:321–329.

Pabo, C. O. and Lewis, M. (1982). *Nature*, 298:4443–447.

Schevitz, R. W., Otwinowski, Z., Joachimiak, A., Lawson, C. L., and Sigler, P. B. (1985). *Nature*, 317:782–786.

Steitz, T. A., Ohlendorf, D. H., McKay, D. B., Anderson, W. F., and Matthews, B. W. (1982). *Proc. Natl. Acad. Sci. USA*, 79:3097–3100.

Sung, M. T. and Dixon, G. T. (1970). *Proc. Natl. Acad. Sci. USA*, 67:1616–1623.

Wohlberger, C., Done, Y., Ptashne, M., and Harrison, S. C. (1988). *Nature*, 335:789–795.

Zubay, G. and Doty, P. (1959). *J. Mol. Biol.*, 1:1–20.

Discussion

Q: Do the different binding modes of the α-turn-α motifs correlate with physical quantities, such as association constants? What is the range of these association constants; is it the same?

A: In all of these proteins the association constants vary with ionic strength. So it depends on what you pick for your standard condition. Depending on your choice, a protein like Cro may even bind RNA or single stranded DNA. The association constants for random sequences or non-specific binding is about 10^7 and that's true of almost all of the repressors and activators — $10^7 - 10^8$. The association constant with operator varies from somewhere between 10^{10} and 10^{11} for Cro and up to about 10^{14} for the Lac Repressor. A correlation between how the helix-turn-helix unit approaches the DNA and the binding affinity is not immediately apparent, but it would be interesting to look more carefully.

Q: You mentioned that you used a fixed fragment length multiple sequence alignment program. Could that have biased you to find a motif that doesn't have insertions, which then locked you into finding or not finding motifs that do have insertions? Has anyone reported a motif with insertions?

A: There are a fair number of structures that we have now and they all have that motif without any insertions. That is a selected group and so you can always say that you are missing the ones that are different. Again, it comes down to a point where you have to say where are you going to make the divisions between this motif and another motif that is related. What would be missed using a fixed fragment length would primarily be proteins with a different length loop or turn between the helices. But this would alter the structure. A simple example is the lexA protein, does that fit into the same motif or does it not? The structure is somewhat different and it may not be any closer to the helix-turn-helix motif than to some of the other proteins that have pairs of helices. One other reason we didn't allow insertions in our initial searches was that we could calculate the statistics more accurately.

Q: Would it be fair to say that the characteristics of the residues making up these sequences is in fact more important than the actual amino acids themselves? For example, could you search with some sort of hydrophobicity string which would give you equal, or perhaps even better results?

A: I'm not sure that the signal to noise would actually get much better. One way you can do that is by changing the similarity table that you use for scoring one amino acid versus another. We've done this using minimum base changes per codon or minimum mutational distance and we've done it with hydrophobicity and other residue characteristic scales and they all work about the same way. What seems to work best depends in part upon what sequence you are testing.

Q: Could I suggest to you an extreme version of what you are saying or seem to be saying — may be implicitly suggesting, and that is that the helix-turn-helix motif has nothing to do with DNA *per se* but is purely a way of getting a protuberance from a globular structure which does not normally occur in other proteins.

A: I wouldn't say that it has nothing to do with DNA *per se* — it has to do with DNA in the sense that an α-helix is nicely complementary to the major groove. The size of that α-helix, even if you rotate it around its axis, is still more or less comparable to the size of the major groove. However, to a certain extent what you suggest is true — it is a way to hold that structure on the surface of the protein where it sticks out enough that it can interact in the major groove. It is how you get your functional site on the outside of the protein, sticking out instead of on the inside in a cleft, where you have it in most enzymes.

Q: Concerning model building, let's just ask the minimal question: Do you think we could model-build the orientation between the two α-helices let alone the interactions with DNA just from the amino acid sequence? In other words, in the last case you said there was a phenylalanine that tended to push them apart. Do you see any hope here at all? Obviously people are going to be anxious to do so.

A: This might be possible with what we know at the present time, but it will be affected by the nonconserved parts of the protein. The second part of the problem, determining the orientation of the recognition helix relative to the major groove, is the central point in modelling how the protein is going to bind and will depend upon a quaternary structure that you can't predict at the present.

The two α-helices themselves, in most of these proteins, are very similar, almost within the errors of the coordinates. In the case of 434 Cro and the 434 Repressor the whole structures are almost identical and it is subtle differences in the interface between the two subunits that alter the way the motif approaches the DNA. Perhaps when we recognize what causes the variation in quaternary structure and you look at all of the structural features, you may be able to model the structures but I don't see being able to do so in the near future.

Q: I just wanted to note that we tested to see whether if you took a hydrophobicity profile and added to it the fact that you had a glycine turn and added to it the fact that at least one of the regions was somewhat basic, that you can in fact get a very complex regular expression. It'll walk through and pick out about 99 per cent of the identified ones and there's only, I think, two false positives that are potential false positives, so you can take that pattern, write it out and add a hydrophybicity profile. Then you're not looking at the amino acids directly at all.

A: Essentially, we're doing that in an indirect way because we make a set of master sequences — we pick as different sequences as we can which we know still have the motif. When you do the sequence comparison you're just finding something common among them. It is a profile which is not defined.

Q: The advantage that was suggested is that it turns out just in terms of speed — you're talking about half a second to search most of the protein database with a regular expression, that's all.

A: Yes.

Q: Is there any evidence that the α-loop-α is stable on its own? Have people looked at peptides in solution to see if there is any evidence for helices or, in the same line, is there any intron/exon evidence that suggests that it is a motif which can be passed round between proteins?

A: These proteins are all prokaryotic proteins and have no introns. The homeodomain proteins from vertebrates may be somewhat similar. They don't fit all of the sequence constraints that you find in the prokaryotic proteins but nonetheless adopt a helix-turn-helix structure. If you count the homeodomain, then yes, you can seem to shuffle them around and get them in many different proteins. There is one other thing that is interesting. If you look at the sequence of the λ Cro and the λ CII protein, within the helix-turn-helix and in the helix which precedes it, you can see a very strong similarity in the DNA base sequences of the genes. Outside that region they are very different, as if that unit was at some point transferred into another context.

Assignment of α-Helices in Multiply Aligned Protein Sequences — Applications to DNA Binding Motifs

Toby J. Gibson

European Molecular Biology Laboratory
Meyerhofstrasse 1
Postfach 102209
D6900 Heidelberg
Germany

gibson@embl.bitnet

Usage of Hydrophobic Information in Helical Prediction

The first solved protein structures were helical globular proteins and soon after their availability for structural analysis, it was noted that buried surfaces of α-helices were composed of hydrophobic residues. Schiffer and Edmundson (1967) introduced the helical wheel representation, in which residues are positioned at 100° intervals around a circle (i.e. 3.6 residues per turn) and suggested its use as a predictive tool for helices. Given the variability of helical length and degree of burial in the tertiary structure, the use of this tool alone proved too simplistic and the presence or absence of helices could not be predicted with a high degree of certainty. Nevertheless, the frequent use of helical wheels to plot segments of sequence with good helical amphipathicity testifies to the continuing utility of this simple representation.

As part of a more complex set of rules for secondary structure prediction, derived partly from an analysis of protein structures and partly from *a priori* expectation, Lim (1974) utilised logic concerning amphipathic periodicity. Although some of the rules employed by Lim may not be well supported by examination of the current database of structures, the overall set of rules works well enough that the method performs as well as other secondary structure prediction methods, being correct to about 60% (Kabsch and Sander, 1983).

A gradual accumulation of knowledge from many sources (too wide to review here) led to the realisation that many membrane-spanning regions of proteins were helical in nature and that they showed strong hydrophobicity for some 20–25 residues. These segments are the most predictable of all secondary structures and plots of, for example, the hydrophobicity index of Kyte and Doolittle (1982) are routinely scrutinised for new protein sequences. Nevertheless, doubts about the number of membrane-spanning helical segments, and even their existence within a protein, regularly occur because ligands from hydrophobic sidechains are important organisers of membrane protein structure and, therefore, occasional membrane-spanning helices are no more hydrophobic than would be a helix nearly buried in a large globular protein. An additional problem is that membrane spanning helices also tolerate proline at a reasonable frequency. Doubts are most often resolved by comparison to other members of the protein family.

Eisenberg *et al.*(1984) provided a widely used method which attempts to quantify the amphipathicity of a sequence in helical (or strand) periodicities. The potential of a sequence to form a half-buried helix is plotted as its hydrophobic moment. While this method undoubtedly detects strongly amphipathic secondary structures, it suffers from a number of drawbacks as a more general predictive tool. Firstly, only a minority of helices are exactly, or even nearly, half-buried and thus, not only do non-helical segments and membrane-spanning helices score equally poorly, but quarter or three-quarter buried helices are not strongly predicted. Secondly, although exposed residues in proteins are on average hydrophilic, they are not, individually, constrained to be so, leading to the not uncommon situation where an exposed but structurally irrelevant hydrophobic residue masks the helical amphipathic periodicity.

Recently, helices have been predicted for a number of small domains by examination of the aligned family of related sequences. The power of this approach has been demonstrated by direct confirmation of the helical assignment in two of these cases. The information which has been used is basically the same as is used in the methods for helical prediction in a single sequence but acquires more certainty when applied to a family of homologous sequences. Nevertheless to date this approach has been used in an *ad hoc* manner and is in need of systematisation. In the next section, examples of helical prediction in DNA-binding domains (which are currently the subject of intense scrutiny) will be examined, after which the rules that were employed will be outlined.

Examples of Helical Prediction in DNA-Binding Domains

Leucine Zipper

Examination of the aligned sequences of the DNA-binding domains from a homologous family of eukaryotic transcriptional activators revealed a heptad periodicity of conserved leucine residues over 28 out of the ~ 60 amino acids in this domain. Given the paucity of proline and glycine in this region, together with experimental evidence that the domain dimerised, Landshutz *et al.*(1988) proposed that the domain formed a single helix and dimerised as a paired antiparallel helical coiled-coil (Crick, 1953). On the basis of experiments which ruled out an antiparallel conformation, O'Shea *et al.*(1989) revised the model to a parallel coiled-coil. The correctness of this model has been confirmed by 2D-NMR studies of synthetic leucine zipper peptides (Oas *et al.*, 1990; Saudek *et al.*, 1990). Attention was also focussed on the second part of the domain, the basic region which binds directly to DNA. Although there are no conserved strongly hydrophobic residues in this domain, there is a conserved residue periodicity with pronounced helical sidedness — for example the least conserved residues are all on the same side on a helical wheel representation. Vinson *et al.*(1989) proposed two additional helices in this region to track around the major groove of DNA. In fact, the periodicity is consistent with one continuous helix (O'Neil *et al.*, 1990), corroboration of which appears in 2D-NMR data that show consecutive peptide NH groups in helical conformation (Saudek *et al.*, 1990). Thus a picture emerges for the DNA-binding domain monomer as essentially a single helix, stabilised by hydrophobic dimerisation contacts over part of its length and DNA-binding contacts in the remainder. Fig. 1 shows the aligned family of sequences found in the SWISSPROT protein sequence database together with the conserved periodic information. The helical prediction, as well as being essentially correct, served as the starting point for the subsequent experimentation, including the choice of peptide and use of 2D-NMR, leading to a good understanding of the domain structure before X-ray crystallographic methods could be applied.

Zinc Finger

Examination of 39 aligned sequences of a 28-residue motif designated the zinc finger (Miller *et al.*, 1985) by both CHOU-FASMAN type prediction and helical wheel representation allowed Brown and Argos (1986) to predict a short helix with the conserved hydrophobic residue periodicity of L..H...H. This prediction was utilised in two tertiary structure model building attempts for the zinc finger (Berg, 1988; Gibson *et al.*, 1988). The solution of zinc finger structure by NMR (Lee *et al.*, 1989) demonstrated the existence of this helix, the length of which proved to be accurately predictable using information from 150 aligned sequences, such as the distribution of allowed Pro and Gly near the helix termini (Gibson *et al.*, 1988). In this example, alternating amphipathic periodicity was also used to assign two short strands (Berg,

```
       1.......10........20........30........40........50........60...
       |        |         |         |         |         |         |
c-fos  ekrriRrerNkmAAakCRnRRreltdtLqaetdqLedeksaLqteianLlkekekLefilaaH..
fos-b  ekrrvRrerNklAAakCRnRRreltdrLqaetdqLeeekaeLeseiaeLqkekerLefvlvaH..
fra-1  errrvRrerNklAAakCRnRRkeltdfLqaetdkLedeksgLqreieeLqkqkerLelvleaH..
c-jun  ikaerKrmrNriAAskCRkRKleriarLeekvktLkaqNseLastanmLreqvaqLkqkvmnH..
jun-d  ikaerKrlrNriAAskCRkRKlerisrLeekvktLksqNteLastaslLreqvaqLkqkvlsH..
jun-b  ikverKrlrNrlAAtkCRkRKleriarLedkvktLkaeNagLssaaglLreqvaqLkqkvmtH..
gcn4   dpaalKrarNteAArrSRaRKlqrmkqLedkveeLlskNyhLenevarLkklvger*
y-AP1  etkqkRtaqNraAqrafReRKerkmkeLekkvqsLesiqqqneveatfLrdqlitLvnelkky..
bzlf1  seleiKrykNrvASrkCRaKfkqllqhyrevaaaksseNdrLrlllkqMcpsldvdsii..
creb   rkrevRlmkNreAAreCRrKKkeyvkcLenrvavLenqNktLieelkaLkdlychksd*
cre-bp1 ekrrkflerNraAAsrCRqKRkvwvqsLekkaedLsslNgqLqsevtlLrnevaqL..
ATF-1  lkreiRlmkNreA.reCRrKKkeyvkcLenrvavLenqNktLieelktLkdlysnksv*
ATF-2  ekrrkvlerNraAAsrCRqKRkvwvqsLekkaedLsslNgqLqsevtlLrnevaqLkqlllah..
ATF-3  erkkrRrerNkiAAakCRnKKkektecLqkesekLesvNaeLkaqieeLknekqhLiymlnlh..
ATF-4  dkklkKmeqNkrAAtryRqKKraeqeaLtgeckeLekkNeaLkeradsLarelqyLkdlieev..
ATF-6  lrrqqRmikNreSAcqSRkKKkeymlgLearlkaalseNeqLkkengrLkrqldevvselrns..
hXBP-1 ekalrRklkNrvAAqtaRdRKkarmseLeqqvvdLeeeNqkLflenqlLrekthgLvvenqel..
c/ebp  neyrvRrerNniAvrkSRdKakqrnvetqqkvleLtsdNdrLrkrveqLsreldtLrgifrqL..
TGA1a  ekvlrRlaqNreAArkSRlRKkayvqqLensklkLiqleqeLerarkqgmcv..
TGA1b  ekkraRlvrNreSAqlSRqRKkhyveeLedkvriMhstiqdLnakvayIiaenatLktq*
HBP-1  lkkqkRklsNreSArrSRlRKqaeceeLgqraeaLkseNssLrieldrIkkeyeeLlskntsL..
cys-3  aaeedKrkrNtaASarfRiKKkqreqaLeksakeMsekvtqLegriqaLetenkwLkglvtek..
v-maf  lkqkrRtlkNrgyAqsCRfRRvqqrhvLeseknqLlqqvehLkqeisrLvrerdaykekyekL..
CPC1   dvvamKrarNtlAArkSReRKaqrleeLeakieeLiaerdryknlalahgaste*
opaque2 ervrkRkesNreSArrSRyRKaahlkeLedqvaqLkaeNscLlrriaaLnqkyndanv..

cons   .....+...N+.AA..cR.++...%..L...#..L...n..L...#..l...%..l...%..%
```

FIGURE 1. Aligned domains of the leucine zipper family taken from the SWISSPROT protein sequence database. Residues 1–24 are defined as the basic region while residues 25–60 are defined as the leucine zipper, although the monomer is now known to be a continuous helix. The consensus is given below the sequences. Symbols for conserved properties are: + for positively charged; # for conserved hydrophobicity; % for partially conserved hydrophobicity. Helical periodicity extends over the whole domain either as conserved hydrophobicity or as conserved residue type.

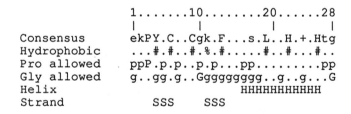

```
                  1.......10........20......28
                  |        |         |        |
Consensus         ekPY.C..Cgk.F...s.L..H.+.Htg
Hydrophobic       ...#.#..#.%.#.....#..#....#..
Pro allowed       ppP.p.p..p.p...pp........pp
Gly allowed       g..gg.g..Ggggggggg..g..g...G
Helix                                HHHHHHHHHHH
Strand            SSS       SSS
```

FIGURE 2. Information derived from 150 aligned zinc finger motifs showing the consensus sequence, conserved hydrophobicity and allowed Pro and Gly positions. The helix and the two strands are confirmed by NMR. The symbol % denotes a conserved hydrophilic amino acid (Lys) which is substituted by a hydrophobic residue in some of the motifs.

1988; Gibson *et al.*, 1988) which were modelled as a hairpin — correctly built by Berg. In Fig. 2, the information on residue conservation and allowance together with observed secondary structure is summarised for the zinc finger.

The Myb DNA-Binding Domain

DNA-binding by Myb utilises a repeated 51–53 amino acid domain. Inspection of the aligned domains from a number of Myb sequences (Fig. 3) shows that about 40 amino acids can be assigned by amphipathic periodicity into 3 consecutive helices with loops defined by frequent Pro and Gly and allowed insertions. Primed by this assignment, it was noticed (Frampton *et al.*, 1989) that helix 2 shared the motif IA..L with helix 2 from another DNA-binding domain, the homeodomain, for which the structure is solved (Billeter *et al.*, 1990). Fig. 3 shows that, when conserved properties are matched, 13 of the 21 positions showing conservation in Myb can be aligned with the homeodomain consensus, allowing insertion to occur only at the known loops, although only 6 of these positions share exact identities. Without the prior assignment of the helices, the inclusion of Myb into the helix-turn-helix class of DNA-binding domains could not have been achieved because the direct sequence similarity to the homeodomains is below detectable levels.

Helix-Loop-Helix Domain

A growing family of DNA-binding domains, originally characterised by its related-ness to a domain in the Myc oncogene, is, notwithstanding the complete absence of direct structural data, now known as the helix-loop-helix domain (Murre *et al.*, 1989). Fig. 4 shows that the aligned sequences reveal two segments with amphipathic helical periodicity (and lacking Pro and with rare Gly) interrupted by a region which tolerates the occurrence of both Pro and insertions. Because the domain is known to dimerise, a schematic model for the domain with four helices packed in longitudinal orientation has been proposed (Davis *et al.*, 1990). This model is remarkably similar to the solved structure of the prokaryotic RNA-binding regulatory protein Rop, a dimeric four-helix bundle (Banner *et al.*, 1987). Indeed, although there are just five apparently conserved amino acids, the similarity of domain size and loop position point strongly to the conclusion that these domains are homologous (Gibson *et al.*, 1991). This conclusion is tentative because the hydrophobic periodicity over the second helix of Rop does not superimpose exactly onto the eukaryotic domain second helix, implying either that the domains are unrelated or that the helix packing has undergone some rotational

```
                    1.......10........20...    .....30.... ....40........50..
                    |        |          |         |          |          |
Mouse-1             lgktrwtreedeklkklveqngt    ddwkvianylp nrtdvqcqhrwqkvlnpe
Chicken-1           lgktrwtreedeklkklveqngt    edwkviasflp nrtdvqcqhrwqkvlnpe
Human-A1            wnrvkwtrdetdklkklveqhgt    ddwtliashlq nrsdfqcqhrwqkvlnpe
Human-B1            kckvkwtheeteqlralvrqfgq    qdwkflashfp nrtdqqcqyrwlrvlnpd
Drosphila-1         gfgkrwsksedvllkqlvethg     enweiigphfk drleqqvqqrwakvlnpe
Maize-1             vkrgawtskeddalaayvkahge    gkwrevpqkaglrrcgkscrlrwlnylrpn
Mouse-2             likgpwtkeedqrviklvqkygp    krwsviakhlk grigkqcrerwhnhlnpe
Chicken-2           likgpwtkeedqrvielvqkygp    krwsviakhlk grigkqcrerwhnhlnpe
Human-B2            lvkgpwtkeedqkvielvkkygt    kqwtliakhlk grlgrqcrerwhnhlnpe
Drosophila-2        likgpwtrdeddmviklvrnfgp    kkwtliaryln grigkqcrerwhnhlnpn
Maize-2             irrgnisydeedliirlhrllg     nrwsliagrlp grtdneiknywnstlgrr
Mouse-3             vkktswteeedriiyqahkrlg     nrwaeiakllp grtdnaiknhwnstmrrk
Chicken-3           vkktswteeedriiyqahkrlg     nrwaeiakllp grtdnaiknhwnstmrrk
Drosophila-3        ikktawtekedeiiyqahlelg     nqwakiakrlp grtdnaiknhwnstmrrk


Predicted Helices        hhhHHHHHHHHH        hhHHHHHHHH     hhhHHHHHHHHH
Hydrophobic         %....#.......#..#%..%..    ..#..##.%#.  ..%...#...#...#...
Allowed pro         ....p..................    p......pp..p  ..............p.
Myb consensus       #.k..WT.eEd..#..##...G     ^..W..IA..l. ^Rt..q#...W...l..$
Myb/HD consensus    ..+..#T..$...#...#....     ^.....IA..L. ^.t..q#...#......
HD consensus        ++.R..yt..q...L...F....yl....r..iA..L. lte.qik.WFqnrR.k.Kk
HD allowed pro      ppp..p..pp...........'.p.p..p........... .pp..............
HD helices                HHHHHHHHHH          HHHHHHHHHH     HHHHHHHHHH
```

FIGURE 3. Aligned Myb repeats from chicken, mouse, human and maize genes. Allowed Pro, hydrophobic conservation, derived helices and the consensus are also shown. For the homeodomain (HD) is shown a consensus from 109 proteins in the SWISSPROT database, allowed Pro and the helices assigned by NMR. The Myb/HD consensus is derived by matching the Myb and the HD consensus together. Symbols: # for conserved hydrophobicity; % for partially conserved hydrophobicity; $ for either Glu or Gln; ^ for insertion allowed in Myb domains.

readjustment. If the domains are related then the initial helical assignment will have again lead to the actual structure before it has been directly demonstrated.

Rules for Detection of Helices in Multiply Aligned Sequences

The above examples illustrate the kinds of rules which have been used successfully to assign helices. It is noteworthy that in all cases, the helices were detected and assigned by visual inspection of aligned sequences. In no cases were automatic methods used to assign the helices (although they were of course used for further justification if they concurred with the visual inspection). The processes used to assign the helices have not been implemented in computer programs.

The importance of aligned sequences is that both conserved and allowed residues are detectable whereas in a single sequence, all residues get treated equally. Additionally, insertion/deletion points are powerful markers for loops. It is possible to ask both "where can a helix occur?" and "where can a helix not occur?". Helices were immediately ruled out in the above examples when any of the following occurred, usually in combination: frequent Pro and/or Gly; allowed insertions; lack of residue or property conservation for 5 or more consecutive positions. They were also ruled out if there was some conservation but the periodicity could not be matched to a helix — as in the β-strand prediction in the zinc finger motif. Having ruled out helices for certain segments, any remaining segments can then be examined for the ability to positively assign a helix by amphipathic periodicity.

The most powerful evidence for the presence of a helix is a periodicity of hydrophobic residues in the well known heptad repeat (#...#..#). However this definition is much too restrictive and is not matched by several of the above examples. It works very well for parallel coiled-coil (Cohen and Parry, 1990) but in general the degree of burial and the length of the helix affect the conservation pattern. Examination of Figs. 1–4 reveals considerable variation in the patterns of hydrophobic conservation for the example helices. A periodicity of conserved hydrophilic residues can also be indicative of a helix, as in the basic regions of both the leucine zipper and helix-loop-helix domains, where residues on one side of the helix would be available to make ionic contact to the DNA backbone. Conserved hydrophilic residues may also be buried within a protein structure if they provide hydrogen bonds, salt bridges, ligands to either metal ions or cofactors, or if they act as catalytic residues, most of which are quite buried. The

```
              1.........10........20......              30........40........50......58

h-tal     ..vrriftnsrerwrqqnvngafaelrkli         pthpdkklskneilrlamkyinflaklln..
h-Lyl1    ..arrvftnsrerwrqqnvngafaelrkll         pthpdrklsknevlrlamkyigflvrllr.
m-myoG    ..drraatlrekrrlkkvneafealkrst          llnpnqrlpkveilrhaiqyierlqalls.
m-MyoD    ..drrkaatmrerrlskvneafetlkrct          ssnpqrlpkveilrnairyieglqallr.
m-Myf5    ..drrkaatmrerrlkkvnqafetlkrct          ttnpnqrlpkveilrnairyieslqellr.
xl-Twst   ..sqrvmanvrerqrtqslneafsslrkii         ptlpsdklskiqtlklasryidflcqvlq.
dm-Twst   ..ngrvmanvrerqrtqslndafkslqqii         ptlpsdklskiqtlklatryidflcrmls.
dm-Da     ..errqannarerlrirdinealkelgrmc         mthlksdkpqtklgilnmavevimtleqqvr.
h-tfe3    ..qkkdnhnlierrrfnindrikelgtli          pkssdpqmrwnkgtilkasvdyirklqkeqq.
h-E12     ..errvannarerlrvrdineafkelgrmc         qlhlnsekpqtklllhqavsvilnleqqvr.
h-E47     ..errmanarerrvrdineafkelgrmc           qmhlksdkaqtklllilqgavqvilgleqqvr.
h-cMyc    ..vkrrthnvlerqrrnelkrsffalrdqi         pelennekapkvvilkatayilsvqaeeq.
h-nMyc    ..errnhnilerqrrndlrssfltlrdhv          pelvnekaakvvilkateyvhslqaeeh.
h-lMyc    ..tkrknhnflerkrrndlrsrflalrdqv         ptlascskapkvvilskaleylqalvgaek.
zm-arlc   ..qirinhvssekkrreleraifdelvavv         psihrvnkasilaetiaylkelqrrvq.
sc-Ino4   ..dkreshkhaeqarrnrlavalhelasli         pdlqpqesrseliiylkslsylswlyerne..
sc-Pho4   ..drrsnkpimekrrrarinnclnektli          paewkgqnvsaapskattveaacryirhlqqngst*
dm-Hairy  ..ylkvkkplleqrrarmnkcldtlktli          ldatkkdparnsklekadilektvkhlqelqrqqa..
dm-Esm5   ..yrkvmkpllerkrrarinkcldelkdlm         aefqgddalirmdkaemleaalvfmrkqvvkqq..
dm-Esm7   ..yrkvmkpllerkrrarinkcldelkdlm         aecvaqtgdakfekadilevtvqhlrklkeskk..
dm-Esm8   ..yqkvkkpmlerqrrarmnkcldnlktlv         aelrgddgilrmdkaemlesavifmrqqktpkk..
dm-Ast3   .psvarrnarenrvkqvnngfvnlrqhlpqtvvnslsn  ggrgsskklskvdtlriaveyirglqdmld.
dm-Ast4   .qsvqrrnarenrvkqvnnsfarlrqhipqsiitdltk  gggrgphkkiskvdtlriaveyirslqdlvd.
dm-Ast5   .psvirrnarenrvkqvnngfsqlrqhipaaviadlsngrrgigpgqanklskvstlkmaveyirrlqkvlh.
dm-ast8   .qavarrnarenrvkqvnngfallrekipeevseafeaq gagrgaskklskvetlrmaveyirslekllg..

Pro       p....P.......%......p.........p......    p.pppppp.ppp....p.....
Hphobic   .....%....%...%#...#....#..%.%..%.##..##.#....
Cons      ......Er.R......f..L.......         ...K..L.a..Yi..l....
-"-          r    l L                            e    Y    l
Shared    *mtkqektalnmarfirsqtltlleklneld        adeqadiceslhdhadelyrsclarfgddgenl*
Rop       #...##.##.#.##.#.#.#.#                  #...#.#%.%#..#%.##.#....#
Hphobic   
Helices   HHHHHHHHHHHHHHHHHHHHHHHHHHHHH           HHHHHHHHHHHHHHHHHHHHHHHHHH
```

FIGURE 4. Aligned domains of the helix-loop-helix family and the Rop RNA-binding protein. Structurally relevant information is given below the sequences: Pro – positions were Pro is allowed; Hphobic – Hydrophobic positions; Cons – residue conservation; Shared – residues present in HLHs and Rop; Rop – the Rop amino acid sequence; Helices – the helical residues in the Rop structure.

conservation pattern may not fit exactly to a repeat of 7 but may approximate to 6 or 8 — e.g. the second helix of the HLH domain cannot be fitted to an exact periodicity but varies within this range. Since proteins are not neat geometrical shapes, but irregular objects, the possibility exists that the buried helical side changes sides, making it difficult on amphipathic grounds alone to predict the helix. An example of this occurs in helix 2 of the steroid receptor DNA-binding domain (Härd *et al.*, 1990) which has the following conserved amino acid pattern C..CR#.KC#.# — the N-terminal cysteines are part of a Cys/zinc cluster on the opposite side of the helix with respect to all the other conserved hydrophobic residues.

Hydrophobic periodicity is a reflection that helices are not stable unless packing interactions occur which, most commonly, are confined to one side of the helix. However to take account of cases like the steroid receptor, a flexible treatment of conserved periodicity is required. It appears that a helix can be allowed if the periodicity of residue conservation is consistent with a face of the helix making packing interactions and may be extended into a region with a different periodicity provided that the helix is not terminated on other grounds.

It has been noted that the end turns of helices show strongly skewed amino acid occurrences relative to the average helical occurrence (Argos and Palau, 1982; Richardson and Richardson, 1988). This information, applied to sequence alignments, ought to be useful in assigning the ends of helices although examination of the example helices would show that they do not all conform to expectation. The two strongest items of information are the favourability of Pro in the first two helical positions and the 20–30% occurrence of Gly at the C-terminal cap position. The latter information is especially variable among helices — both the zinc finger helix and Myb helix 1 are preferentially capped by Gly while no other sample helix shows a Gly preference. This observation therefore cannot be applied as part of a statistical ensemble for C-terminal assignment since it would then discriminate against some two-thirds of helix C-termini. The presence of a conserved Gly adjacent to a helical prediction may indicate a helix cap but its absence does not rule it out.

Assignment of β-Strands

This type of logic should be applicable to the remaining regular secondary structure of proteins, that is to say the prediction of β-strands, as illustrated by the identification of the two β-strands in the zinc finger (Berg, 1988; Gibson *et al.*, 1988). However, the rules will be more complicated. Alternating periodicity will only work for strands packing on one side and exposed on the other. This is usually not true for parallel β-sheet which is buried on both sides. Also, strands which form the edges of a sheet are not properly buried on the hydrophobic contact side and can show extensive hydrophilicity as noted by Lim (1974) in his secondary structure prediction algorithm, as well as being rather tolerant of Pro and occasionally exhibiting insertions. β-bulges (Richardson *et al.*, 1978) disrupt the amphipathic periodicity and may lead to confusion with helix, although the allowance of Pro and Gly in the bulge should overrule the helix.

Automation

The successes of visual inspection in the assignment of helices mean that it is highly desirable to automate helical prediction from aligned sequences. The main reason why this has not been done already is that, to develop and test an algorithm requires a database of aligned families of sequences for which at least one member of the family has a solved structure. Such databases are now being prepared (Sander and Schneider, 1990) and when they achieve a sufficient reliability of sequence alignment, the success rate of the rules outlined above and the evaluation of new rules may be undertaken.

It may be worth pointing out that visual inspection mainly utilises Boolean logic and not the type of statistical analysis used by most secondary structure prediction algorithms. For example very often there is a two state model for hydrophobicity: either a residue can be treated as a buried hydrophobic or it cannot — whereas in most analyses it has been usual to create an index of relative hydrophobicities of the amino acids — necessitating the immediate use of statistical preferences that seem to lead inexorably to the 60% cutoff barrier for secondary structure prediction.

The successes of the visual inspection method have centred on small domains. This is mainly because they are simple enough to be manageable, but it should be borne in mind that the rules are almost certainly weaker in large globular proteins. For example, frequent Gly is a good indicator of loops in the example DNA-binding domains because it is so strongly excluded from the example helices, occurring at 1.2% in the leucine zipper, 0.7% in the zinc finger, 0.4% in the Myb repeat and 2.1% in the HLH domain. Thus the frequency of Gly in the example helices from small domains is below the average frequency for helices in globular proteins, 5.0% (Chou and Fasman, 1974). Similarly there are a number of examples of conserved Pro in kinked helices. Large proteins are expected to be better able to compensate for the energetic penalty represented by such deviations from the ideal. Therefore rules for secondary structure prediction from multiple sequence alignments will not be able to yield 100% correct predictions. However, in addition to offering the prospect of an improvement over the 60% boundary of existing approaches, the lessons so far indicate that they are capable of predicting some segments of protein families to a reliability approaching 100%.

References

Argos, P. and Palau, J. (1982). *Int. J. Protein Peptide Res.*, 19:380–393.

Banner, D. W., Kokkinidis, M., and Tsernoglou, D. (1987). *J. Mol. Biol.*, 196:657–675.

Berg, J. (1988). *Proc. Natl. Acad. Sci. USA,*, 85:99–102.

Billeter, M., Qian, Y. Q., Otting, G., Müller, M., Gehring, W. J., and Wüthrich, K. (1990). *J. Mol. Biol.*, 214:183–197.

Brown, R. S. and Argos, P. (1986). *Nature*, 324:215.

Chou, P. Y. and Fasman, G. D. (1974). *Biochemistry*, 13:211–222.

Cohen, C. and Parry, D. A. D. (1990). *Proteins*, 7:1–15.

Crick, F. H. C. (1953). *Acta Crystallogr.*, 6:689–697.

Davis, R. L., Cheng, P. F., Lassar, A. B., and Weintraub, H. (1990). *Cell*, 60:733–746.

Eisenberg, D., Weiss, R. M., and Terwilliger, T. C. (1984). *Proc. Natl. Acad. Sci. USA*, 81:140–144.

Frampton, J., Leutz, A., Gibson, T. J., and Graf, T. (1989). *Nature*, 342:134.

Gibson, T. J., Postma, J. P. M., Brown, R. S., and Argos, P. (1988). *Protein Eng.*, 2:209–218.

Gibson, T. J., Sibbald, P. R., and Rice, P. (1991). In press.

Härd, T., Kellenbach, E., Boelens, R., Maler, B. A., Dahlman, K., Freedman, L. P., Carlstedt-Duke, J., Yamamoto, K. R., Gustafsson, J.-Å., and Kaptein, R. (1990). *Science*, 249:157–160.

Kabsch, W. and Sander, C. (1983). *FEBS Letts.*, 155:179–182.

Kyte, J. and Doolittle, R. F. (1982). *J. Mol. biol.*, 157:105–132.

Landshutz, W. H., Johnson, P. F., and McKnight, S. L. (1988). *Science*, 240:1759–1764.

110

Lee, M. S., Gippert, G. P., Soman, K. V., Case, D. A., and Wright, P. E. (1989). *Science*, 245:635–637.

Lim, V. I. (1974). *J. Mol. Biol.*, 88:873–894.

Miller, J., McLachlan, A. D., and Klug, A. (1985). *EMBO J.*, 4:1609–1614.

Murre, C., Schonleber McCaw, P., and Baltimore, P. (1989). *Cell*, 56:777–783.

Oas, T. G., McIntosh, L. P., O'Shea, E. K., Dahlquist, F. W., and Kim, P. S. (1990). *Biochemistry*, 29:2891–2894.

O'Neil, K. T., Hoess, R. H., and DeGrado, W. F. (1990). *Science*, 249:774–778.

O'Shea, E. K., Rutkowski, R., Stafford, W. F., and Kim, P. S. (1989). *Science*, 245:646–648.

Richardson, J. S., Getzoff, E. D., and Richardson, D. C. (1978). *Proc. Natl. Acad. Sci. USA*, 75:2574–2578.

Richardson, J. S. and Richardson, D. C. (1988). *Science*, 240:1648–1652.

Sander, C. and Schneider, R. (1990). *Proteins*, 8:3.

Saudek, V., Pastore, A., Castiglione Morelli, M. A., Frank, R., Gausepohl, H., Gibson, T., Weih, F., and Roesch, P. (1990). *Protein Engineering*. In press.

Schiffer, M. and Edmundson, A. B. (1967). *Biophysical J.*, 7:121–135.

Vinson, C. R., Sigler, P. B., and McKnight, S. L. (1989). *Science*, 246:911–916.

An Expert System for Secondary Structure Prediction

S.R. Presnell[1], B. I. Cohen[1] and F. E. Cohen[1,2]

Departments of Pharmaceutical Chemistry[1] and Medicine[2]
University of California, San Francisco
San Francisco, CA 94143-0446
U.S.A.

cohen@cgl.ucsf.edu

Introduction

The problem of aligning protein sequences would disappear if we could solve the protein folding problem. The three-dimensional structures could be compared to decide if two proteins were related. Unfortunately, there seems to be little hope of solving the problem in the near future. This chapter will focus on sequence patterns within primary structure which correlate with secondary structure (Kabsch and Sander, 1984). Protein chain folding is likely to involve significant structural feedback. Primary structure is not the sole determinant of secondary structure. As secondary structure forms, there are implications for the formation of other secondary structures, so the issues of context-dependent secondary structure and tertiary structure interactions will need to be addressed.

We would like to be able to take a structure like flavodoxin, and recognise the sequence patterns which stabilize the tertiary structure, in this case, hydrophobic patches accross the surface of helix charge pairs with a helical periodicity and a cluster of hydrophilic residues in the loop regions (Cohen *et al.*, 1982). Reading through the sequence, it is possible to recognize these patterns and their associated three-dimensional structure (See Fig. 1). To the extent that this analysis is generalisable, then secondary structure prediction is possible. At some level, this process is secondary structure prediction by homology. The structural homology between helices translates into sequence similarity. Of course, at this remote level of similarity, any evolutionary connection to the original structure and the original sequence is lost.

At the first level one could look at some sequence and assign propensities about an individual residue's tendency to form helical structure, beta structure or turns. This was the approach of Chou and Fasman (1974). The problem is that this very local

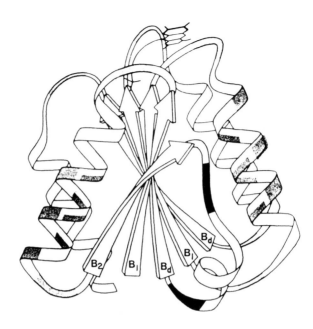

FIGURE 1. Schematic diagram of flavodoxin (Smith *et al.*, 1977) with some structural patterns highlighted. Most patterns are left out for clarity.

compositional information gives no insight into the folding of the chain. It reveals the relative importance of local structure in chain folding. Predictions based on this type of information consistently attain 60% to 65% accuracy (Fasman, 1989; Schultz, 1988). Unfortunately, there seems to be a barrier to improving this result if only local information is used. The logical way to overcome this barrier is to devise methods to incorporate long range information.

Proteins whose structure is dominated by α-helices provide a convenient starting point for a sequence structure analysis. Haemoglobin is an example of an all helical molecule (see Fig. 2). This molecule can be constructed from a series of isolated helices which pack together to form a tertiary structure. By identifying the helical residues and learning the rules for helix packing, the structure of haemoglobin, or, presumably by analogy, myoglobin or any number of other all-helical proteins would become obvious. This presupposes that the known protein structures reveal information about protein organisation. Rules deduced from the analysis of crystallographic data could be used to predict protein folding.

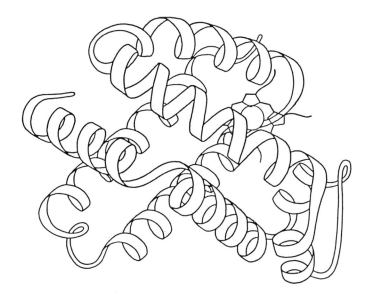

FIGURE 2. Schematic diagram of hemoglobin (after Richardson)

Four Helix Bundles

Four helix bundles are an interesting subset of all helical proteins with extensive molecular interactions between the helices. The extent of the interactions between two helices can be quantified as the change in accessible surface area between the isolated and interdigitated conformations (Richmond and Richards, 1978). It became clear from this analysis that hydrophobic residues tend to be at the centre of these interactions and, furthermore, that the shape of the organisation of hydrophobic was related to the inter-helical packing angle (Cohen *et al.*, 1979). We define a four helix bundle as a collection of four helices such that each helix buries a surface greater than 10% of the total molecular surface area. To be consistent with the work of Weber and Salemme (1980), the packing angle between any two helices was constrained to be less than 40 degrees. Four helical bundles can have zero, one, two or three overhand connections. These structures could be left handed or right handed (see Fig. 3) (Presnell and Cohen, 1989).

If one took four helices and calculated all possible organisations, 48 would be possible. If the anti-parallel alignment of dipoles is important (Hol *et al.*, 1979), then only six structures are sensible. If no overhand connections are made then the arrangement shown in the upper left corner of Fig. 4 is the only right-handed structure

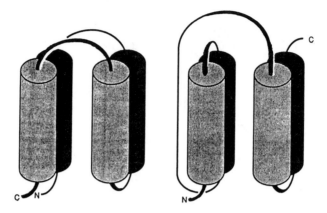

FIGURE 3. Two left-handed bundles (side view). Three specific attributes fully describe the topology of a four-α-helix bundle. These are (i) the polypeptide backbone connectivity between helices, (ii) the unit direction vectors of the individual helices, and (iii) the bundle handedness. In the first bundle there are no overhand connections, and in the second bundle there is one overhand connection. The handedness of a particular bundle is determined using the "right hand rule" of physics. To determine if a helix bundle is of a particular handedness, orient the thumb of one hand parallel to the first helix or helix A where the positive unit vector stems from N terminus to C terminus. Helix B should be orientated to the left if it is a left-handed bundle or to the right if it is a right-handed bundle. In the case where helix B is diagonally opposed to helix A, the handedness is then based on the position of C helix relative to helices A and B.

allowed and its topological enantiomer is the only left-handed structure possible. If two overhand connections are allowed, the last two plausible structures are specified. This analysis presumes that the helix dipole is important.

Analysing the database of structures, one can find four-helix bundles as isolated domains. They also occur as a subdomain or super-secondary structure within a larger protein (see Table 1). It was suggested by Weber and Salemme (1980) that all four-helix bundles would be right-handed and have no overhand connections. However, we find there is an equal number, if not more, left-hand four-helical bundles (Presnell and Cohen, 1989) and most have no overhand connections. It has been argued, based on the economy of chain tracing, that there should never be any overhand connections. In fact there are some examples that have one overhand connection, both in left and right-handed form; and one example with two overhand connections. However, we have yet to find a right-handed structure with two overhand connections but it is logical to expect that such a structure exists. Cytochrome P450 was the only structure that did not conform with the topological restrictions expected of four-helix bundles.

The original distribution of 48 possibilities would have predicted that 7 structures should not fit the complementary dipole hypothesis for every one that does. The ob-

Right-handed all anti-parallel bundles:

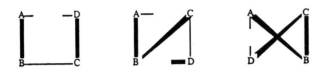

Left-handed all anti-parallel bundles:

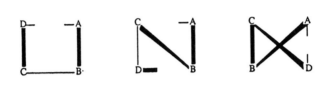

FIGURE 4. Schematic representation of the possible anti-parallel four-α–helix bundles (top view). Bold lines represent connections in front of the page; thin lines represent connections behind the page. Left-handed and right-handed forms of four-α–helix bundles have an equal probability of occurrence.

Topologies of currently known four-α-helix bundles

Overhand connection(s), no.	All anti-parallel		Others (right-handed)
	Left-handed	Right-handed	
0	Complement C3a	Cytochrome b-562	
	Complement C5a	Cytochrome c'	
	Cytochrome b_5	Methemerythrin	
	Interleukin 2	TMV coat protein	
	T4 lysozyme		
1	Ferritin	Phospholipase C (b)	CytochromeP-450$_{cam}$
2	Human growth hormone		

There are no left-handed topologies for "other" four-α-helix bundles. TMV, tobacco mosaic virus.

TABLE 1.

116

served distribution is quite asymmetric (1:13) and suggests that the dipole moment must have some importance in chain folding. Economical connections between consecutive helices cannot be a critical factor since 5 out of 14 structures occur with multiple overhand connections. The challenge is to learn how to put these long-range effects within proteins to use in predicting structure.

Secondary Structure Prediction

The ability to recognise sequence-structure correlations can be codified as a set of sequence patterns with structural implications (e.g. Cohen *et al.*, 1986). Initial efforts have been directed at the set of all-helical proteins. A statistically valid approach to protein recognition requires the division of all helical proteins into two sets — a development set, which can be examined to develop patterns, and a test set, which is used only for testing. One of the problems with patterns developed on one set of proteins is that the patterns may reflect the peculiarities of the proteins studied, instead of the underlying principles of protein structure. The test set helps avoid this problem and differences in accuracy between the development and test sets quantify the degree to which the pattern simply reflects the development set.

At UCSF, Abarbanel developed PLANS, a regular expression formalism for describing sequence patterns in proteins and nucleic acids (Abarbanel, 1984). One of us, B.I. Cohen, has enhanced this tool by creating a window-based, mouse driven pattern matcher that facilitates queries of a database of protein sequences and annotated secondary structure information, providing both visual and numeric feedback. We have added a feature to this program which allows us to identify a marker (e.g. turns) which can be used to parse the structure into subregions. For example; subregions bounded by turns should contain, at most, one unit of secondary structure. It should be noted that these regions are allowed to overlap, reflecting the difficulty in specifying the exact beginning and end of a substructural feature.

Following the work of Rose (1978) we define turns as positions where the chain changes direction. Typically this occurs between units of secondary structure. In globular proteins, turns are solvent exposed segments of the chain, evenly distributed through the sequence, and dominated by hydrophilic residues. Sometimes turns can be located as weakly hydrophilic regions appropriately spaced between other stronger turns. The chain tends to start on one side, traverse the centre of the molecule as a secondary structure segment and then chain direction. In an all-helical protein, the diameter of

FIGURE 5. Schematic representation of a polypeptide chain constrained to adopt a globular structure. The proximity of turns to the surface is highlighted by the shaded circles.

particle (from its molecular weight) divided by the pitch of the helix (1.5Å/residue) yields the maximum spacing between turns. An algorithm which would allow the placement of consecutive turns at, say, residues 20 and 60 could not be guaranteed to produce a globular protein.

The combination of hydrophilicity and the appropriate spacing of turns through the sequence is very powerful. The length of the sequence between turns is typically 12 residues for beta-strands and about 22 residues for alpha-helices. After finding the "strong" turns, the algorithm examines the interval between these turns. If the intervening region is greater than the allowed spacing between turns, a spatially appropriate region between the turns is searched in greater detail to see if a "weak" turn exists (see Fig. 5). A series of regular expressions have been composed which define turns in a hierarchical way. The simplest patterns are used to help define the intervening patterns guided by the appropriate spacing.

In the development set, 42 out of 47 turns were correctly identified and three known helical regions were broken by turns. The results from the test set are quite comparable: 43 of 51 turns were correctly identified and four helices were split representing a 5 percent decrease in accuracy compared with the development set. This suggests that the turn algorithm, when focused on all- alpha proteins will achieve about 85 percent accuracy by taking advantage of hydrophilicity plus the appropriate spatial distribution of turns. Thus, turns can provide a natural parsing of the structure.

The region of chain bounded by turns can then be examined for any periodicity. Using only all-alpha structures there are no beta-strands to consider which makes the task simpler. However, it is important to recognise that helices are not homogeneous entities, but rather structures that have a distinct start, middle and end. Hydrophobic patches are known to dominate the surface of helices. This was first analysed systematically by Schiffer and Edmonson (1967). Richmond and Richards (1978) exploited this feature to identify helix-helix packing sites, while Eisenberg *et al.*(1984) have employed a hydrophobic moment analysis to classify this structural property. We have recognised that they provide a useful pattern for identifying the middle of alpha-helices. Argos and Padlan (1982) and Richardson and Richardson (1988) analysed the end of helices (N and C caps) and noted the tendency for charged and hydrophilic residues to interact with the N and C-termini of the alpha-helices. Putting all these ideas together, we can begin to recognise the segments defined by the turn patterns as helical or non-helical.

Fig. 6 defines our helical core pattern. The segment: $\phi * * \phi\phi * * \phi$ corresponds to four hydrophobic residues (ϕ) forming one face of the helix ($*$ is any amino acid). Similarly, one can define a charged pair with a sequential spacing of three or, more commonly, four residues as stabilising a helical conformation. An N-cap can be defined as a sequence which facilitates helix termination or favourably interacts with the positive pole of the helix dipole. One useful pattern identified a negatively charged residue followed by a capping residue in phase with the hydrophobic face of the helix. The phasing of the hydrophobic face with the capping residues appears critical (see Fig. 7). C-caps were also examined for residues which could interact with the negative pole of the helix (Fig. 7).

The above patterns were applied to the test set of proteins. 48 of the 55 helices were correctly by the hydrophobic core pattern and one region which is not helical was misclassified. Table 2 shows the results for the capping patterns on both the development and test data sets of proteins. We are currently attempting to combine these patterns with a syntax sensitive to the chain direction to produce a complete secondary structure assignment.

In summary, it seems to be possible to recognise about 60–70 percent of structural features with reasonable certainty using pattern descriptors that can recognise context dependent patterns. The distribution of these features along the chain creates a directional syntax — turn, N-cap, middle, C-cap turn, which appears to improve secondary structure prediction (Fig. 8). For the class of all-helical proteins, it seems possible to

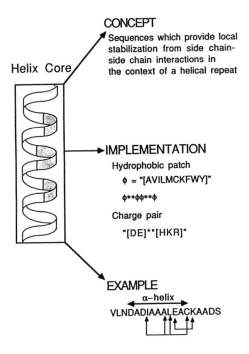

FIGURE 6. Representative patterns for the prediction of helix cores

Secondary Structure Prediction Accuracy - α/α Proteins

	Development Set	Test Set
Turns	89%	84%
Helix Core	87%	82%
N-Cap	75%	65%
C-Cap	74%	58%

TABLE 2.

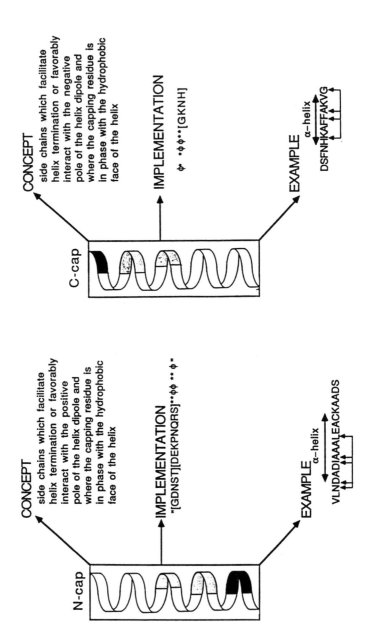

FIGURE 7. Representative patterns for the prediction of helix N-cap (left) and C-cap (right) are shown.

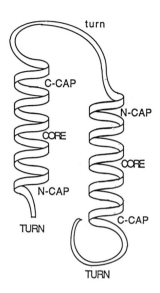

FIGURE 8. Sequential pattern syntax and its structural justification.

create an algorithm which is accurate 80 percent of the time.

Acknowledgements: The authors acknowledge with thanks the support of the NIH (GM39900), the Searle Scholars/Chicago Community Trust, and the Defence Advanced Research Projects Agency under contract N00014-86-K-0757 administered by the office of Naval Research.

References

Abarbanel, R. A. (1984). PhD thesis, University of California, San Francisco.

Argos, P. and Palau, J. (1982). *Int. J. Peptide Res.*, 19:380–393.

Chou, P. Y. and Fasman, G. D. (1974). *Biochemistry*, 13:211–245.

Cohen, F. E., Abarbanel, R. A., Kuntz, I. D., and Fletterick, R. J. (1986). *Biochemistry*, 25:266.

Cohen, F. E., Richmond, T. J., and Richards, F. M. (1979). *J. Mol. Biol.*, 132.

Cohen, F. E., Sternberg, M. J. E., and Taylor, W. R. (1982). *J. Mol. Biol.*, 156:821–862.

Eisenberg, D., Weiss, R. M., and Terwilliger, T. C. (1984). *Proc. Nat. Acad. Sci. USA*, 81:140–144.

Fasman, G. (1989). The development of the prediction of protein structure. In Fasman, G. D., editor, *Prediction of Protein Structure and the Principles of Protein Conformation*, pages 193–316. Plenum, New York.

Ferrin, T., Huang, C., Jarvis, L., and Langridge, R. (1988). *J. Mol. Graphics*, 6:13–37.

Hol, W. G. J., Van Duijnen, P. T., , and Beresden, H. J. C. (1979). *Nature*, 273:443–446.

Kabsch, W. and Sander, C. (1984). *Proc. Nat. Acad. Sci. USA*, 81:1075–1078.

Lim, V. I. (1974). *J. Mol. Biol.*, 88:857–894.

Presnell, S. R. and Cohen, F. E. (1989). *Proc. Nat. Acad. Sci. USA*, 86:6592–6596.

Richardson, J. R. and Richardson, D. C. (1988). *Science*, 240:1648–1652.

Richmond, T. J. and Richards, F. M. (1978). *J. Mol. Biol.*, 119:537–555.

Rose, G. D. (1978). *Nature(London)*, 272:586–590.

Schiffer, M. and Edmundson, S. B. (1967). *Biophys. J.*, 7:121–135.

Smith, W. W., Burnett, R. M., Darling, G. D., and Ludwig, M. L. (1977). *J. Mol. Biol.*, 117:195.

Taylor, W. R. (1988). *Protein Engineering*, 2:77–86.

Weber, P. and Salemme, F. R. (1980). *Nature (London)*, 287:82–84.

Discussion

Q: Isn't glycine useful as an initiator or terminator of α-helices?

A: The N- and C-cap patterns shown were only part of a longer list of capping patterns. The statistics in Table 2 dealt with our best guess of all possible capping patterns. A difficult aspect of this work is that glycine can sit in the middle of helices as well as at the end of helices. Glycine seems to stabilize helix-helix interactions in the globin family (Richmond and Richards, 1978; Cohen *et al.*, 1979).

Q: Should proline have a more prominent role in initiating or terminating helices?

A: Yes, but it is in 10 percent of alpha helices. These realities complicate prediction.

Q: What about the role of charged residues at the termini of helices interacting with the helix dipole?

A: This information was useful but not helpful enough.

Q: Did the set of proteins studied include only four helix bundles?

A: The data set that we were looking at for secondary structure prediction did not include just helix bundles — there were a number of other proteins that are exclusively helical.

Q: What was the degree of homology between the sequences in the development and test sets for secondary structure prediction? Is the work of Lim (1974) instructive?

A: We did our best to try and pick sequences that were relatively unrelated. The limited number of all-helical proteins complicates this task. It is interesting to note that there seem to be two groups of all-helical proteins, those that have long helices and several of them, e.g. four, and those that form box-like structures. The rules seem to be slightly different. Efforts were made to construct generic, physically reasonable patterns. You are correct to point out Lim's seminal contribution.Certainly, much of our work is based on his insightful paper in 1974. Unfortunately, it is almost impossible to produce a rigorous algorithm which recreates Lim's work because of internal inconsistencies.

Q: Doesn't the test set contain useful information for pattern development?

A: With this approach, one only has a single chance to be blind to the test set. Once uncovered,the data set is no longer pristine. We tried to pick a test set that was relatively different than the development set. I think that the fact that the accuracies decrease by 5 percent reflects the fact that there was at least a 5 percent learning effect. There are also other all-helical structures that were not used because their coordinates have not been published and those will be good tests.

Q: Is there a four helix bundle in thermolysin?

A: Yes, but the packing angle in thermolysin is too wide to meet our initial criteria. However there is a very interesting group of structures that have a wide packing angle. I believe they are also four helix bundles — strictly, it is a continuum between relatively parallel and perpendicular arrangements.

Q: Why was an angular restriction placed on helix-helix packing in four bundles?

A: The cut off was chosen to mime what the literature (Weber and Salemme, 1980) suggested a four helix bundle should be. Clearly, that needs to be extended. It is a structural continuum and an arbitrary cutoff was necessary to begin a taxonomic survey.

Q: In general, one does not know protein class. What can be done in this case?

A: There are three distinct possibilities. Circular dichroism spectroscopy can assess secondary structure content. The second possibility is that three or four algorithms are applied to each sequence. When these secondary structures are assembled, it may be clear that not all the tertiary structures are possible. The third possibility is to develop an algorithm which helps recognize protein class. Such algorithms have been developed and their accuracy is approximately 80 percent. It is certainly true that when one focuses on one protein class it becomes easier to write these patterns. It is also true that the value of these patterns degrades as other protein classes are examined. When the all beta turn algorithm is applied to an all alpha sequence the protein is carved into segments appropriate for beta structure. The intervening segments are not typical of beta strands. This approach was explored as a method for determining protein class. To date, this has not proved successful.

Q: Can the four helix bundle criteria be altered so that the packing dictated by the helix dipole is always observed?

A: The one bad example would go away along with the two good examples. Cytochrome P450 appears to create a problem by adopting this unusual topology. Presumably, residues from other parts of the chain satisfy the electronic constraints. The more chain you have, the more ways you have of avoiding an inherent difficulty using some other piece of chain.

Patterns in Secondary Structure Packing — a Database for Prediction

Nigel P. Brown

Laboratory of Mathematical Biology
National Institute for Medical Research
The Ridgeway, Mill Hill
London NW7 1AA
U.K.

n_brown@uk.ac.mrc.nimr

Introduction

Most known structures have been derived by X-ray crystallographic means. This technique requires good quality crystals of the protein and almost all structures are of globular proteins. Because of the difficulty of crystallising membrane bound proteins, there is a paucity of structures for these (Eisenberg, 1984). 2D NMR is a relatively new technique producing distance or angle constraints which may be used to determine structures for small proteins in solution (polypeptides of up to around 100 residues have been attempted). Both methods have the practical limitation of requiring samples of the material in appropriate form and purity.

By comparison, prediction approaches which rely on the rapidly expanding libraries of sequence data might be able to bypass this problem. However, energy minimisation methods are computationally intensive and suffer from explosive combinatorial complexity as the number of atoms and therefore the number of interactions to be modelled increases. A particular problem is that of modelling the interaction with solvent. Only structures of order tens of residues have been predicted successfully. Pattern directed prediction methods seem to offer the best hope for structure determination as techniques exist for reducing some of the inherent combinatorial complexity in the pattern generation and application stages (for example dynamic programming). The other techniques then have complementary roles in structure validation and structure refinement.

Prediction approaches have been influenced by the hierarchical model of protein structure (see Schulz and Schirmer, 1979; Jaenicke, 1987), which comprises the primary, secondary, tertiary, and quaternary structural levels. Of course, every level need not be present in a given protein: Many proteins are monomeric and thus lack a quaternary level, or they may consist of more than one structural domain giving a complex tertiary organisation. The scheme breaks the structure prediction problem into a set of sub-problems of lesser complexity, providing a simple framework for techniques working at each level.

As a first approximation, the model is useful as shown by the partial success of secondary structure prediction methods (see for example von Heijne, 1987; Taylor, 1988) and, at the other extreme, in predicting quaternary structure from the known surface characteristics of the monomer (Schulz and Schirmer, 1979), as the structure of the monomer is only relatively slightly changed on aggregation. Amino acid substitution studies (Bowie *et al.*, 1990), and sequence alignments in protein families (Zvelebil *et al.*, 1987) reveal the presence of feedback from higher levels, as many different sequences correspond to highly similar structures. They also show that certain residues (mostly hydrophobic core) in the sequence are highly conserved, indicating the structural importance of these 'scaffold' positions.

Any successful prediction method must take these factors into account (the many to one mapping of sequence to structure, and the importance of scaffold positions), and will build on knowledge of both sequence and structural patterns. Patterns and models at the supersecondary and structural domain/tertiary levels of structure will be addressed below.

Topology, Packing and Geometry

Topological models, for both supersecondary and domain structure, derive from analysis of sequentially connected secondary structures along the polypeptide chain. Differentiating factors include the number of participating secondary structures, choice of parallel or anti-parallel pairwise orientation, handedness, and connectivity. Many examples at both structural levels have been described (Schulz and Schirmer, 1979; Richardson, 1981; Richardson, 1985). Most of these structures have been recognised by visual study of the structure databanks, although formal means of describing and searching for known topological motifs have also been developed using declarative techniques (Rawlings *et al.*, 1985) and databases (Islam and Sternberg, 1989; Gray *et al.*, 1990).

A different approach has been used to characterise close packed structures by analysis of packing areas and residue properties, and three-dimensional geometry. These analyses concentrate on physicochemical and spatial interactions, often ignoring connectivity attributes.

Estimates of packing area can be derived by comparison of the accessible surface area (Richards, 1977) of structural segments in unpacked and packed states (Richmond and Richards, 1978). Residue properties such as hydrophobicity and size have also been studied in relation to location in the interface region between packed structures. As a first approximation, helices and sheet elements may be represented simply by axial vectors whose pairwise orientations are described by distances and angles. Models for helix-helix, helix-sheet, and sheet-sheet packing based on observations of these attributes have been proposed (reviewed in Chothia, 1984).

Important properties of a method of structure description that allows structure comparison are internal consistency and completeness: the parameters must be independent of the original coordinate system in which they were derived, and must be sufficient to reconstruct the structure at that level of resolution.

A natural extension of the geometry approach is to consider all pairwise arrangements in a domain, rather than just the close packing pairs, as in the work of Richards and Kundrot (1988), who developed six pairwise geometric parameters, such as distance and interaxial angle. Visual representations of some of the parameters were demonstrated in the form of 'distance' plots of a sequence against itself, wherein the parameter (or parameters) of interest were shown in the cells of the matrix. A scheme for searching for homologous substructures by comparing distance matrices containing multiple parameters was outlined but not elaborated.

The description method of Richards and Kundrot is overdetermined in that the complete arrangement of vectors can be reconstructed from various combinations of the pairwise information given the sequence ordering. It is, however, a very rich representational scheme. Other workers have used a more economical system of parameters to model structure. Abagyan and Maiorov (1988) represent regular secondary structure by axial vectors as usual, but represent also the intervening loop regions by vectors linking the endpoints of consecutive structures. The continuous chain of vectors so formed is then characterised by successive vector lengths, vector-vector angles, and torsion angles to give a uniquely determined description of the path of the polypeptide backbone. Some structure comparisons of small $\alpha + \beta$ class proteins are then performed by superimposition of the respective vector chains by a RMSD (root mean square deviation) method (see Orengo, in this Vol.).

A quite different approach to describing tertiary structure is that of Finkelstein and Ptitsyn (1987), who present a general model of protein architecture and attempt to show how this is consistent with various observed properties, such as average secondary structure lengths and orientations (parallel or anti-parallel), and thermodynamic considerations. Treating regular secondary structural elements as rigid rods with some permitted curvature, and ignoring sequence connectivity, globular protein structural domains are thought of as layers or stacks of rods. These are of four types: flat monolayer, rolled, quasi-spherical or multiple-layered.

Continuing in this vein, Murzin and Finkelstein (1988) show that the geometry of many proteins consisting of α-helices packed around a ball-like core can be described by quasi-spherical polyhedra in which the core is formed by the helix packings. The polyhedron forms a *virtual framework* for the polypeptide chain such that (some) edges of the polyhedron correspond to helices, and the order (i.e. number of faces) of the polyhedron depends upon the number of helices to be packed about the core. The precise layout of the helices is combinatorially limited and further constrained by simple rules. For example, the chain cannot pass inside the framework, and no vertex can be visited more than once, i.e. the backbone cannot cross its own path.

Parameter Selection

The work presented here seeks to further explore packing geometry and topological relationships in more detail than many previous studies given the faster computing hardware and larger amounts of on-line storage now available. This will form a basis for identification of new structural motifs and for development of new folding algorithms.

The initial stage required the generation of a databank of geometrical relationships and accessible areas lost on packing for α and β secondary structure pairings using an objective definition of secondary structure, DSSP (Kabsch and Sander, 1983). The basic scheme is outlined in Fig. 1.

For the geometry calculations, helical and sheet elements were represented by axial vectors derived from the coordinates of participating main chain Cα, C, N, and O atoms of the secondary structure. Each vector was derived by analysis of the moments of inertia of these points, treating them as unit masses (Taylor *et al.*, 1983), the smallest moment giving the axial vector, with direction defined from N-terminus to C-terminus of the structure. The endpoints of each axis were defined by the projections of the end Cα atom coordinates onto this vector.

The relative orientations of pairwise vectors were measured in terms of distances

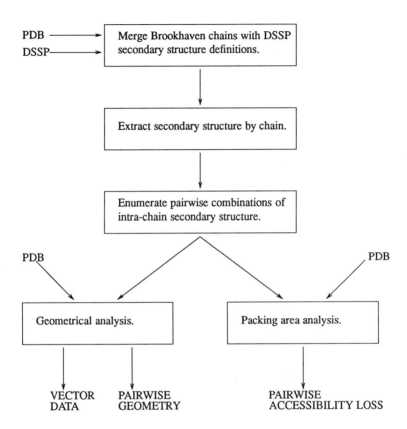

FIGURE 1. Stages in the generation of axial vectors, pairwise geometry, and pairwise packing of secondary structures. Inputs are Brookhaven PDB chain definitions and coordinates and DSSP structure definitions.

Protein structure coordinates were taken from the Brookhaven Protein Data Bank (PDB), EMBL release, July 1989. Secondary structure definitions were taken from the Definition of Protein Secondary Structure (DSSP) databank (Kabsch and Sander, 1983) of the same release, but with the PDB chain definitions supplanting those of DSSP. Pairwise combinations of these structures for a given protein chain were generated and analysed to examine geometrical relationships and loss of accessible surface area on packing. Accessibilities were estimated as the difference between the total accessible surface for the isolated structures and the pair *in situ* using the ACCESS program after Lee and Richards (1971).

ACCESS was ported to UNIX by the author from a version at Birkbeck College written in FORTRAN 66 under Dec VMS. Other programs to generate the vectors and pairwise geometry were written in C linked with routines from the EISPACK Fortran 77 library from Argonne National Labs. All code runs under SunOS4.0 UNIX on Sun-3 or Sun-4 platforms.

and angles between the vectors when projected into a plane. A projection is defined by a line connecting the (non-parallel) vectors and a plane normal to this, into which the vectors are projected. The length of the connecting line is taken as the distance d between the vectors, while the angle between the vectors in the projection plane defines the interaxial angle Ω of the secondary structures. A second angle τ describes the tilt of the vectors away from the projection plane, see Fig. 2.

Various authors have used different projection schemes, as shown in Fig. 3, which shows three alternative projections. That derived from the common perpendicular CP_d of the two vectors is the most interesting mathematically, as it has the unique property that $\tau \equiv 0$ (Richards and Kundrot, 1988). However, we are interested in the spatial orientation of structures lying on these vectors so that other projections with respect to the line segments are perhaps more useful. The two described here are derived from the closest approach distance CA_d (Chothia et al., 1977; Chothia et al., 1981), which is obtained by projecting each line segment endpoint onto the other vector and selecting that projection which gives the shortest distance, and the inter-centroid distance IC_d which connects the midpoints of the line segments (equivalent to the inertial centroids of the constructor atoms). Data for all three projections were gathered for comparison.

The interaxial angle has been defined in different ways also. Chothia et al. (1977) projected their vector pair into the common contact plane (closest approach projection) and ignored the sense of the vectors to yield an angle in the range $\pm90°$. They used the sign convention such that the angle is negative if the near vector is rotated clockwise with respect to the far vector. This is essentially a torsion angle about the projection plane normal, but limited to $\pm90°$ by ignoring vector sense. In their later paper, Chothia et al. (1981) measured some angles over the full range $\pm180°$, to accommodate features of their helix-helix packing model. Richards and Kundrot (1988) however, used the common perpendicular projection plane and considered vector sense to yield angles over the full range $\pm180°$ with the same sign convention. This angle corresponds exactly to the dihedral angle about the plane normal (common perpendicular).

In this work, various properties, both for single structure data and pairwise data, were analysed with a view to finding a minimal set of parameters that best describe the single vector and pairwise vector property spaces. For the single vector data, these include analysis of distributions and correlations for structure length, helical pitch and various hydrophobic moments. For the pairwise material, distributions and correlations for solvent accessibility and geometry were examined. In particular the various definitions of interaxial distance and angle (see Fig. 3) were compared. All interaxial angles, irrespective of projection, took account of vector sense, were defined

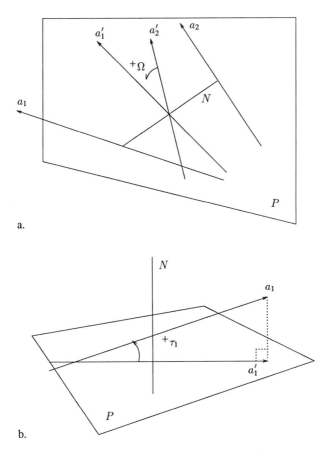

a.

b.

FIGURE 2. Illustration of a projection plane P and the normal N connecting a pair of axial vectors a_1, a_2.

    a. Interaxial angle Ω between projections a'_1, a'_2 in P. If vector sense is retained $|\Omega| \leq 180°$, otherwise $|\Omega| \leq 90°$. The interaxial distance d is the length of N.

    b. Projection of a_1 onto its image a'_1 in P showing τ_1, the associated tilt angle. The pairwise tilt angle τ is the sum $\tau_1 + \tau_2$.

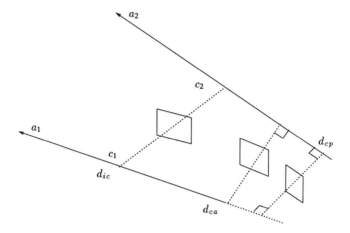

FIGURE 3. Illustration of three contruction lines joining axial line segments a_1, a_2 to yield different projection planes.

CP_d is the common perpendicular of the corresponding *vectors* through a_1, a_2; CA_d corresponds to the closest approach distance when the endpoints of CP_d lie outside one or both of a_1, a_2, and is is the minimal projection of each line segment endpoint onto the other line segment; IC_d corresponds to the inter-centroid distance, i.e. the line joining the midpoints c_1, c_2 of the line segments.

and used the same sign convention as above. Where necessary, conversion to the $pm90°$ scheme of Chothia *et al.* (1977) was readily performed. Various subclasses of the data were examined, such as structure type (DSSP: extended strand (e), 3_{10}-helix (g), α-helix (h) (Kabsch and Sander, 1983)), packed and non-packed structures, and consecutive and non-consecutive structures. The material cannot be presented fully here so some selected results are given.

For example, in comparing the three projection planes (Fig. 3) it was found that the inter-centroid plane yielded the best separation of structural types by distance. Fig. 4a shows the frequency distributions for inter-centroid distance, IC_d, by structural type. The cumulative distribution is also drawn and shows three prominent peaks. The first at about 4.6Å (from examination of a finer resolution distribution, not shown) is almost entirely due to close packed β-strand pairs in sheets and hairpins (compare with Fig. 4b), while the others are cumulative. The extended strand pair curve, (ee), reveals a periodic succession of diminishing minor peaks which probably correspond to successively removed β-strands in sheets. The helix pair curve, (hh), is also of interest as it shows a peak at about 11Å which indicates close packed helices, verified in Fig. 4b.

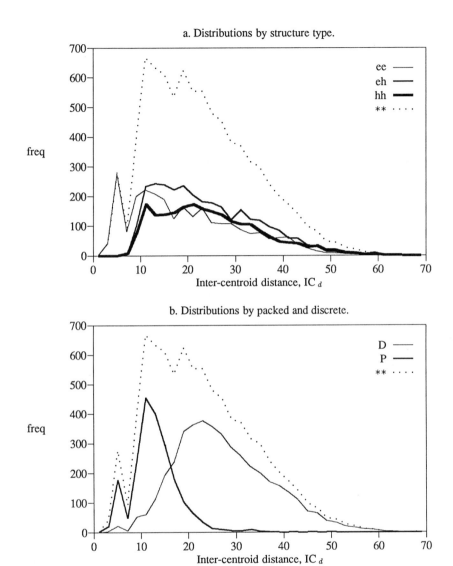

a. Distributions by structure type.

b. Distributions by packed and discrete.

FIGURE 4. Frequency distributions of inter-centroid distance, IC_d (Å), for 8410 secondary structure pairs ($**$) from 119 protein chains, classwidth 2 Å. Minimum structure length is 4 residues. Secondary structure definitions follow those of DSSP (Kabsch and Sander, 1983) and centroids are as defined in the main text.

a. The contributions due to interactions between extended strand (ee, 2645 pairs), extended strand and helix (eh, 2950 pairs), and helices (hh, 2177 pairs) are shown.

b. Contributions due to packed structures (P, 2068) and discrete structures (D, 4263), where packed is defined as ACCESS accessibility loss > 0.1 Å^2.

Fig. 5 shows a polar plot of, this time, common perpendicular distance, CP_d, against the interaxial angle at the common perpendicular CP_{180} (the subscript indicates the choice of $\pm 180°$ angle convention), for extended strand pairs only. Each plotted point lies at the end of a radial vector whose angle about the origin represents CP_{180} and whose length represents CP_d. Points in the right half of the plot indicate parallel pairs, while those in the left half indicate anti-parallel. Again, the periodic separation of β-strands is visible out to 2 or 3 levels on either side of the origin.

Viewed *across* the strands, i.e. edge on, β-sheets have a left-handed twist, so that as the pairwise strands become more remote from each other, the CP_{180} angle will become more negative. The average twist varies with the structural context of the sheet as shown by the following dihedral angle measurements: $-17°$/strand in aligned β-sandwiches (Chothia and Janin, 1981), $-19°$/strand in α/β proteins (Janin and Chothia, 1980), $-24°$/strand in orthogonal β-sandwiches (Chothia et al., 1981). From distributions similar to those in Fig. 4, the mean CP_d for neighbouring β-strands is about 4.6Å. In the figure 'ideal' points for a sheet with average twist $-20°$ $[(-17 - 19 - 24)/3]$ are superimposed on the original data and show a correlation with the concentric clusters.

In addition, there is an obvious directional 'smearing' of the data, out from these loci, which is probably due in part to the large variability of sheet twist and seems to be skewed in favour of sheets with higher than average twist. Inter-sheet interactions between strands in the β-sandwiches mentioned above will also complicate the picture.

Applications and Continuing Work

Continuing with the analytical theme, identifying a minimal vector (of size N) of parameters that best characterise pairwise structures would define an N-dimensional space within which helix-helix, helix-strand, and strand-strand pairs could be plotted. The organisation and clustering of points in this space would be of interest both globally and with respect to protein families. Such a minimal vector has been derived using Principal Components Analysis (PCA) (Chatfield and Collins, 1989), a statistical technique used to transform a set of potentially correlated variables (the geometric parameters in this case) into a (hopefully smaller) set of uncorrelated random variables.

Although a few obvious features have been demonstrated in the simple pairwise analysis described above, no new features have been found, which is not really surprising. However, these simple geometric parameters provide a crude means of comparing protein structures at the tertiary level. The Secondary Structure Alignment Program

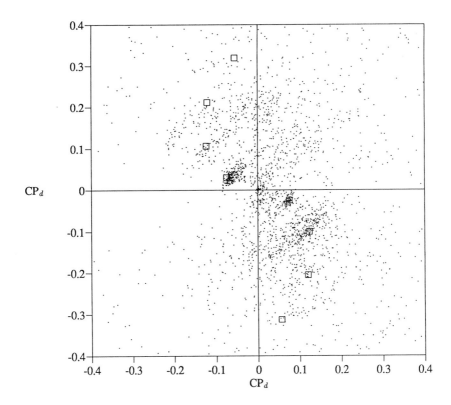

FIGURE 5. Polar plot of common perpendicular distance, CP $_d$ (Å), against common perpendicular angle, CP$_{180}$, for 2645 extended strand pairs (Some peripheral points have been clipped). Points in the right half ($-90°$ < CP$_{180}$ ≤ $+90°$) represent pairs with parallel geometry, while pairs in the left half ($+90°$ < CP$_{180}$ ≤ $-90°$) have anti-parallel geometry. Superimposed on the raw data are some 'ideal' points (boxes) — see main text for explanation.

Minimum structure length is 4 residues. Secondary structure definitions follow those of DSSP (Kabsch and Sander, 1983) and vectors and angle convention are as defined in the main text.

(SSAP) of Taylor and Orengo (1989) (see also Orengo, in this Vol.), compares structures at the residue level to generate an alignment and associated score or goodness of fit. This is an elegant extension of the now conventional sequence alignment techniques which use a form of dynamic programming first applied to biological sequence comparison by Needleman and Wunsch (1970). However, a complete pairwise analysis of the Brookhaven databank against itself would be impracticable at this level because of the time required. By comparing structures at the cruder level of their component secondary structure a much reduced subset of interesting alignments may be be offered for residue level alignment and this has now been achieved for the whole Brookhaven databank (Orengo and Taylor, 1990).

This application of the vector geometry can also be used to identify new motifs in the structure databank by similar means. Work is also progressing on re-examining the original data in terms of 3-way and higher complexity interactions based on the simple pairwise material as this should provide another route to motif characterisation.

References

Abagyan, R. A. and Maiorov, V. N. (1988). A simple qualitative representation of polypeptide chain folds: Comparison of protein tertiary structures. *J. Biomol. Structure & Dynamics*, 5(6):1267–1279.

Bowie, J. U., Reidhaar-Olson, J. F., Lim, W. A., and Sauer, R. T. (1990). Deciphering the message in protein sequences: Tolerance to amino acid substitutions. *Science*, 247:1306–1310.

Chatfield, C. and Collins, A. J. (1989). *Introduction to Multivariate Analysis*. Chapman and Hall.

Chothia, C. and Janin, J. (1981). Relative orientation of close-packed β-sheets in proteins. *Proc. Natl. Acad. Sci. USA*, 78(7):4146–4150.

Chothia, C., Levitt, M., and Richardson, D. (1977). Structure of proteins: Packing of α-helices and pleated sheets. *Proc. Natl. Acad. Sci. USA*, 74:4130–4134.

Chothia, C., Levitt, M., and Richardson, D. (1981). Helix to helix packing in proteins. *J. Mol. Biol.*, 145:215–250.

Chothia, C. (1984). Principles that determine the structure of proteins. *Ann. Rev. Biochem.*, 53:537–572.

Eisenberg, D. (1984). Three-dimensional structure of membrane and surface proteins. *Ann. Rev. Biochem.*, 53:595–623.

Finkelstein, A. V. and Ptitsyn, O. B. (1987). Why do globular proteins fit the limited set of folding patterns? *Prog. Biophys. Mol. Biol.*, 50:171–190.

Gray, P., Paton, N., Kemp, G., and Fothergill, J. E. (1990). An object-oriented database for protein structure analysis. *Prot. Eng.*, 3(4):235–243.

Islam, S. A. and Sternberg, M. J. E. (1989). A relational database of protein structures designed for flexible enquiries about conformation. *Prot. Eng.*, 2(6):431–442.

Jaenicke, R. (1987). Folding and association of proteins. *Prog. Biophys. Mol. Biol.*, 49(2/3):117–237.

Janin, J. and Chothia, C. (1980). Packing of α-helices onto β-pleated sheets and the anatomy of α/β proteins. *J. Mol. Biol.*, 143:95–128.

Kabsch, W. and Sander, C. (1983). Dictionary of protein secondary structure: Pattern recognition of hydrogen-bonded and geometrical features. *Biopolymers*, 22:2577–2637.

Lee, B. and Richards, F. M. (1971). The interpretation of protein structures: Estimation of static accessibility. *J. Mol. Biol.*, 55:379–40.

Murzin, A. G. and Finkelstein, A. V. (1988). General architecture of the α-helical globule. *J. Mol. Biol.*, 204:749–769.

Needleman, S. B. and Wunsch, C. D. (1970). A general method applicable to the search for similarities in the amino acid sequence of two proteins. *J. Mol. Biol.*, 48:443–453.

Orengo, C. A. and Taylor, W. R. (1990). A rapid method of protein structure alignment. *J. Theor. Biol.*, 147:517–551.

Rawlings, C. J., Taylor, W. R., Nyakairu, J., Fox, J., and Sternberg, M. J. E. (1985). Reasoning about protein topology using the logic programming language PROLOG. *J. Mol. Graph.*, 3(4):151–157.

Richards, F. M. and Kundrot, C. E. (1988). Identification of structural motifs from protein coordinate data: Secondary structure and first level supersecondary structure. *PROTEINS: Structure, Function, & Genetics*, 3:71–84.

Richards, F. M. (1977). Areas, volumes, packing, and protein structure. *Ann. Rev. Biophys. Bioeng.*, 6:151–176.

Richardson, J. S. (1981). The anatomy and taxonomy of protein structure. *Adv. Prot. Chem.*, 34:167–339.

Richardson, J. S. (1985). Describing patterns of protein tertiary structure. *Methods Enzymol.*, 115:341–358.

Richmond, T. J. and Richards, F. M. (1978). Packing of α-helices: Geometrical constraints and contact areas. *J. Mol. Biol.*, 119:537–555.

140

Schulz, G. E. and Schirmer, R. H. (1979). *Principles of Protein Structure*. Springer-Verlag.

Taylor, W. R. and Orengo, C. (1989). Protein structure alignment. *J. Mol. Biol.*, 208:1–22.

Taylor, W. R., Thornton, J. M., and Turnell, W. G. (1983). An ellipsoidal approximation of protein shape. *J. Mol. Graph.*, 1(2):30–38.

Taylor, W. R. (1988). Pattern matching methods in protein sequence comparison and structure prediction. *Prot. Eng.*, 2(2):77–86.

von Heijne, G. (1987). *Sequence Analysis in Molecular Biology (Treasure Trove or Trivial Pursuit)*. Academic Press Inc.

Zvelebil, M. J., Barton, G. J., Taylor, W. R., and Sternberg, M. J. E. (1987). Prediction of protein secondary structure and active sites using the alignment of homologous sequences. *J. Mol. Biol.*, 195:957–961.

Secondary and Supersecondary Motifs in Protein Structures

J. M. Thornton[1], B. L. Sibanda[1], C. M. Wilmot[2] and J. Singh[1]

Crystallography Department
Birkbeck College
Malet Street
London WC1E 7HX
U.K.

thornton@uk.ac.ucl.bioc.bsm

Protein Structure Database Development

Our approach to extracting and understanding protein sequence and structural motifs is to combine information derived from protein structures and homologous protein sequence families. Initially however, I will describe our protein database, which has involved a collaboration between Birkbeck College (London) and Leeds University. We have aimed to develop an integrated database of protein sequence and structure, called ISIS (Akrigg *et al.*, 1988). The sequence database was developed at Leeds University principally by John Wootton and Alan Bleasby. Here the main goal was to create a composite sequence database, by amalgamating all the publicly available databases and eliminating duplications (Bleasby and Wootton, 1990). The current version contains about 25,000 sequences. They have also written software for pattern matching using weight matrices and are developing a table of features.

On the structural side, our task at Birkbeck has been to take the known structures from the Brookhaven Data Bank (Bernstein *et al.*, 1977) and derive data, such as ϕ,ψ angles, accessibility, and other information which is useful to have available immediately without having to re-compute it every time. These data are stored and accessed using ORACLE^{TM}, a commercial relational database management system (Islam and Sternberg, 1989). We have found that using this standard relational system, there are definite drawbacks — the principal one being that relational databases are

---

[1] Current address: Biomolecular Structure and Modelling Unit, Biochemistry and Molecular Biology Dept., University College, Gower Street, London WC1E 6BT

[2] Department of Molecular Biology, Research Institute of Scripps, Scripps Clinic, 10666 North Torrey Pines Road, La Jolla, C.A. 92037

not ideally suited to deal with the sequential nature of proteins. This leads to serious problems because the majority of searches are sequential.

To avoid this problem, we are developing an in-house database management system, called IDITIS, which employs Order Based SQL (Structured Query Language). The shell 3-D SCAN includes additional funtionality like template matching within it (Thornton and Gardner, 1989).

The data we include are categorised into different levels: protein, chain, residue, atom, secondary structure and interaction level. (The latter includes hydrogen bonds, salt bridges, disulphide bridges, etc.). We have tables for the secondary structures and are working on motif tables and the homologous protein tables. (Along the way, we have learnt that it is really quite an arduous task to set up complete tables accurately and to get everything within them right.) All our data are derived from Brookhaven coordinate files. (Bernstein *et al.*, 1977). Additionally — we have a side chain database (Singh and Thornton, 1990). The power of these databases will be through the integration of the features section. (A feature can be derived from either the sequence or the structure.)

Protein Structure Hierarchy

We hope to use our database to abstract structural motifs. Fig. 1 represents a very simply hierarchy of one way to describe protein structure. There are several different structural levels, starting with the secondary structure (the α-helices and the β-strands) which in turn form super-secondary structures and then tertiary structures. The super-secondary structures occur across all sorts of homologous families and therefore, they are fundamental building blocks (as important as the amino acids themselves). A major problem in structure prediction is that if there is no known structure of a homologous protein, a reliable prediction cannot be made. To try to improve structure prediction it may help to predict not only the standard α-helix and β-strand, but also the supersecondary structures, like the $\alpha\alpha$-hairpin, $\beta\alpha\beta$-unit and the $\beta\beta$-hairpin.

Over the last several years, we have been looking in detail at these secondary and supersecondary structures. The methodology used in this work, is to look at all known protein structures, (which is why we need the protein structure database), and extract all the information on these different motifs. As an example, I will describe tight turns, called the β-turns.

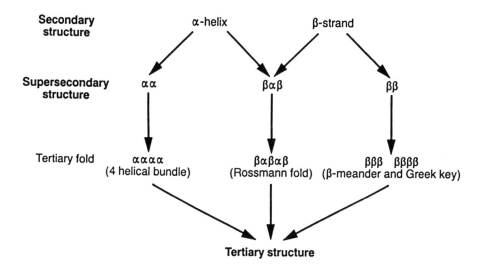

FIGURE 1. Hierachy in protein structure

Analysis of β-Turns

These are specific structural turns (as distinct from wider loops) in which the separation between $C\alpha_i - C\alpha_{i+3}$ is less than 7Å, forming a very tight turn. There must also be the requirement to exclude α-helical residues because they also fulfil this definition. So within the database environment, we have extracted all examples of known turns from proteins solved to high resolution (≤ 2.5 Å).(Wilmot and Thornton, 1988)

There are different types of turns which all satisfy the above criteria, in some the NH of the central peptide points forward in some (type I), while in others (type II), the carbonyl points forward (see Fig. 2). Essentially this is a flip of the central peptide unit. In type II, residue $i + 2$ adopts the α_L conformation on the Ramachandran plot, suggesting that glycine, asparagine or aspartic acid would be preferred. Our hypothesis was that there would be distinct sequence differences between the different types of turns and that previous prediction work, which amalgamated the turns, would lead to loss of information.

We therefore took the turns and classified them into their different types. Here I will highlight a few results. Firstly, the old way of classifying turns, in terms of a defined range of ϕ,ψ angles of the central two residues, is too restrictive. Using those constraints, almost a third of turns are not classified. Inspection of these turns shows

144

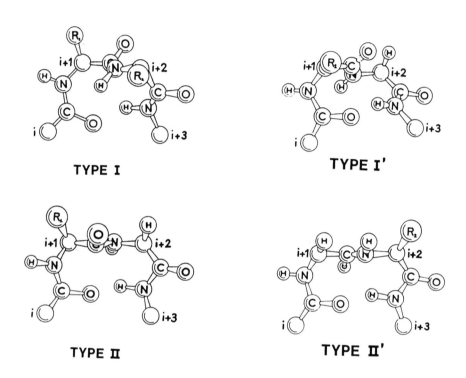

FIGURE 2. Different types of β-turns

β-turn type	Position $l+1$		Position $l+2$		Ramachandran nomenclature[a]	Number of turns located[b]	Number of distorted turns located[c]
	ϕ (°)	ψ (°)	ϕ (°)	ψ (°)			
I	−60	−30	−90	0	$\alpha_R - \alpha_R$	309	119
I'	60	30	90	0	$\alpha_L - \gamma_L$	35	16
II	−60	120	80	0	$\beta - \gamma_L$	113	99
II'	60	−120	−80	0	$\epsilon - \alpha_R$	20	5
VIa	−60	120	−90	0	$\beta - \alpha_R$	3	0
VIb	−120	120	−60	0	$\beta - \alpha_R$	2	1
VIII	−60	−30	−120	120	$\alpha_R - \beta$	62	54
						(394)	100
Total						938	394

[a]Ramachandran regions as given by Thornton et al. (1988).
[b]Turns located within the ϕ, ψ ranges of Lewis et al. (1973).
[c]The 394 type IV turns from the previous column divided into β-turn types, based on the turns adopting the Ramachandran nomenclature of the ideal β-turns.

TABLE 1. Standard β-turn types located.

that they are just distortions of the basic turn types (see Table 1). The second problem is that the current nomenclature I, II, I' etc is difficult to remember and non-informative and this is a handicap.

We therefore classified turns using the conformation of residues ($i+1$ and $i+2$) by reference to regions of the ϕ - ψ plot (see Fig. 3). For example in a type I turn both $i+1$ and $i+2$ adopt an α_R conformation, thus a type I turn becomes an $\alpha\alpha$-turn. A type II becomes a $\beta\alpha_L$-turn. This is a much more logical classification. In our analysis we also found a new turn which we originally called type VIII but in the new nomenclature becomes an $\alpha\beta$turn (Wilmot and Thornton, 1990).

The results of this analysis have been used to improve structure prediction, using the method of Chou and Fasman (1977). We looked at the four positions of different types of turns and calculated which residues are preferred. As expected, a distinct difference was found between the $\alpha\alpha$-turns and the $\beta\alpha_L$-turns — (the type I and the type II). Interestingly, in the $\alpha\alpha$-turns there is a very strong preference for asparagine, serine and threonine at position i. When examined structurally (see Fig. 4) we found that hydrogen bonds often occur between the side chain oxygen of residue i and main chain NH of residue $i+2$. Thus the β-turn is stabilised not only by the standard carbonyl O–NH bond but also by this extra side chain to main chain hydrogen bond. Therefore the preference for asparagine, threonine and serine can be simply explained in terms of residues that are capable of making a hydrogen bond with the NH group (Wilmot and Thornton, 1988).

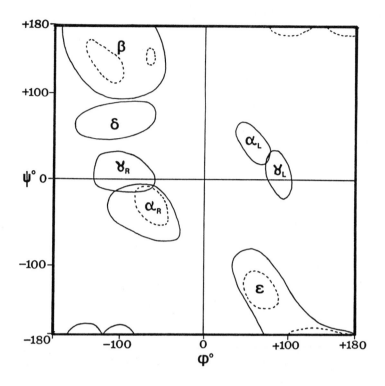

FIGURE 3. ϕ,ψ nomenclature

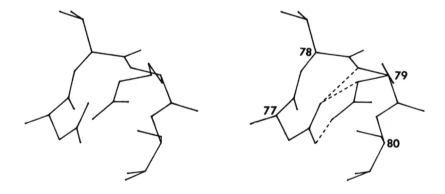

FIGURE 4. An example of Type I β-turn stabilised by an additional hydrogen bond between the side chain of residue i and peptide amide of residue $i + 2$.

To aid prediction we made simple sequence templates from these structures, taking the most preferred residue types at each position. However, searching the sequence database gave many incorrect matches, since the templates were not sufficiently specific. For improved prediction we used a simple multiplication method (Wilmot and Thornton, 1990). Alternatively, a neural network approach, can be used, where correlated preferences can be taken into account (McGregor *et al.*, 1987).

Supersecondary Structures

Supersecondary motifs are widespread within all proteins. This has nothing to do with homology; they are just basic structural units that are common. This occurs because sequential secondary structures, inevitably come close together spatially simply through the dynamics of folding of the chain. We felt that the way forward from conventional structure prediction which involves looking for α-helices, β-sheets and turns, was to look for these supersecondary structures. We began by developing a prediction technique based on lengths of strands and helices combined with secondary structure prediction (Taylor and Thornton, 1984). We have since looked at these structures in more detail to see whether there were any more conserved features that would be useful predictively. Strands and helices are fairly standard, with amphipathic hydrophobic features. So we concentrated instead on the loop regions between these patterns to see if there was any information there or whether, as the conventional wisdom stated, they were just random coils.

We looked therefore for preferred conformations within the loop regions (Thornton *et al.*, 1988). Firstly, we analysed the loop lengths in these different supersecondary structures and what was striking was how short most of these loops were — about 60% had less than six residues (See Fig. 5). We then looked to see if there was any evidence for structural families within the loops. The results are summarised in Fig. 5, where hashed parts represent the members of these supersecondary structures for which the loops fell into one of the recognised structural families. They correspond to the shorter loops, as expected, because with a short loop there is a limited number of ways to achieve loop closure.

Concentrating on loop regions within the β-hairpins, we found a very strong clustering (Sibanda and Thornton, 1985). Fig. 6 shows the ϕ - ψ values of residue $i + 1$ and $i + 2$ in the 2-residue β-hairpins. We found two main conformations which corresponded, as expected, to a turn, but, as we did not expect at all, to the uncommon

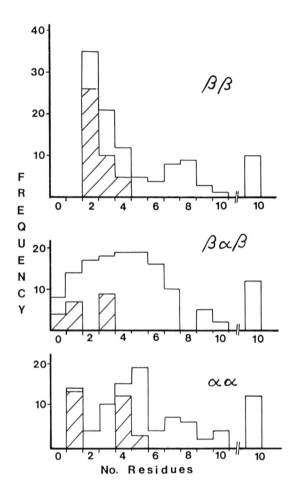

FIGURE 5. Loop lengths in the standard supersecondary structures, β-hairpins, $\alpha\alpha$-loops and $\beta\alpha\beta$-units. The hashed lines denote those structures which adopt one of the common loop structures described in Table 2.

SYSTEMATIC MODELLING OF β-HAIRPINS

—— REPLACEMENT ——→

DELETION ←——————→ INSERTION

SET	DOUBLE H-BOND		SINGLE H-BOND	ALTERNATIVE
A	2:2	Type I' ★ — Gly —Asn—Gly— Asp αL γL; Type II' ★ —Gly—Ser— Thr ε γR; Type I —X—X— αR αR	2:4 unusual — Various	6:6 6:8 10:10 10:12
B +1	3:3	Rare — Various	3:5 Type I [1-4] ★ + G1 β-bulge —X—X—Gly—X— β αR γR γL β; Various	7:7 7:9 11:11 11:13
C +2	4:4	Type I [1-4] ★ —X—X—X—Gly αR αR γR αL; Various	4:6 Many different conformations	8:8 8:10 12:12 12:14
D +3	5:5	Many different conformations	5:7 Many different conformations	9:9 9:11 13:13 13:15

TABLE 2. Classification of hairpins.

150

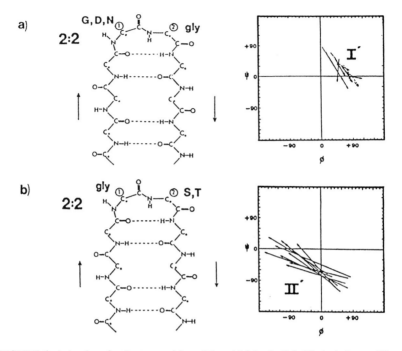

FIGURE 6. ϕ,ψ values for the two residues, L1 and L2 in the 2:2 β-hairpin loops. The ϕ,ψ of residue L1 is indicated by the beginning of an arrow, and L2 by the pointed end.

I' and II' turns. These turns probably occur here because the twist of the turn matches the twist that is always found between adjacent β-strands.

β-Hairpin Classification

There are four different classes of β-hairpins which can be categorised by the numbers of residues in the loop, from two to five residues (see Table 2) (Sibanda *et al.*, 1989; Milner-White and Poet, 1986). Within each class there are two subgroups. The full nomenclature is shown diagrammatically in the table depending on whether or not the distal hydrogen bond is formed (Sibanda *et al.*, 1989).

We looked at the different structures of the examples within these different classes of hairpins and found that many loops include the very tight β-turns that I described above. We found a large family in which the loop conformation involved a type I turn with a G1 β-bulge; and a second family which included a type I turn again within the four residue loop; (Gibson describes yet another family involving cis-prolines in this

Vol.). Table 2 can be used as a simple guide to modelling β-hairpins. For example a two residue loop modified by a one residue insertion, corresponds to a vertical move down the table, e.g. from a 2:2 class to a 3:3 class. Then one can examine the sequence within the loop and ask if it corresponds to one of the standard sequences. For example the 3:5 loop requires a glycine at the fourth position or alternatively it may be one of the rather rare 3:5 hairpins which adopt a variety of other conformations. We have tabulated all of the known examples and this type of data will be available in our database.

β-hairpins with more than 5 residues in the loop are relatively unusual. Within our data set (which was taken from 59 proteins) we had a total of 106 hairpins. We found, surprisingly, that only 30 per cent belonged to this longer loop category.

Because the short loops involve tight β-turns, we built up a sequence pattern for each loop family. For example, the type 2:2 $\alpha_L \gamma_L$-hairpin loop prefers two sequential glycines. However searching in the database for Gly-Gly, there are obviously many examples which do not form β-hairpins. As expected there is a contextual requirement that there must be β-strands on either side of the Gly-Gly turn. It has proved difficult to combine this sort of sequence information with the preferences for forming secondary structures in supersecondary structure prediction and we still really have not solved this problem.

Modelling β-Hairpins

Within a family of homologous proteins, a structurally equivalent hairpin will move between classes as residues in the loops are inserted and deleted (Table 3). For example, in actinidin/papain we have a deletion of four residues within one of the hairpins. In actinidin the hairpin adopts a 2:2 type I' conformation whilst in papain this becomes a 2:2 with a type I turn. In the serine proteinase, there are many examples of residue insertions and deletions in β-hairpin loops. Three such loops are shown in Table 3.

For short hairpins, these sort of rules are very useful. However, they do not apply at all to long hairpins. As is well known, loops mutate wildly at times and because of this variability, the loop regions are the most difficult part to model on the basis of homology. There are many developments in progress at Birkbeck and elsewhere to improve the modelling of these longer loops.

Changes in hairpin classification in homologous proteins

Protein	File	Res. no.†	Sequence	Classification	Ref.
Actindin	2ACT	169	Y G T E <u>G G</u> V D Y	2:2 I′	1
Papain	9PAP	166	Y G <u>P</u> - - - - <u>N</u> Y	2:2 I	2
Elastase	1EST	201	C L V <u>N G</u> Q Y A	2:2 I′	3
S. griseus protease A	2SGA	201	A <u>G</u> - - - - <u>S</u> T	2:2 II′	4
S. griseus protease A	2SGA	48	V S V <u>N G</u> V A H	2:2 I′	4
α-Lytic protease	2ALP	47	V T R <u>G A</u> T K G	2:2 II′	5
γ-Chymotrypsinogen	2GCH	48	N <u>E</u> - - - - <u>N</u> W	2:2 I	6
Elastase	1EST	33	L Q Y R S <u>G S</u> S WA H	2:2 II′	3
γ-Chymotrypsinogen	2GCH	33	L Q <u>D</u> - - - <u>K</u> T G F H	3:5 Iβ	6

References:
1. Baker (1980); 2. Kamphuis *et al.* (1984); 3. Bernstein *et al.* (1977); 4. James *et al.* (1980); 5. Fujinaga *et al.* (1985); 6. Cohen *et al.* (1981).
  These data were extracted using BIPED, the Protein Structure Database developed at Birkbeck supported by the Protein Engineering Club.
  † Residue number of 1st residue given in sequence.

TABLE 3. Changes in hairpin classification in homologous proteins.

β-Arches

So far I have only considered β-hairpins which by definition involve strands which are hydrogen bonded together. However, in the β-sandwich family of protein structures there are also many examples of what we call the β-arches. These involve two strands lying in different sheets on opposite sides of the β-sandwich. They are not hydrogen-bonded together but are connected by a loop arch. (see Fig. 7). We were interested to see if there was any difference in the connecting loop length of the arches when compared to the length of the loops in hairpins. The distributions of loop length, plotted in Fig. 8 are obviously different, showing that, in general, the arches need more residues in the loop than the hairpins. This sort of information can be used predictively since a short loop sequence implies a hairpin, while the arch requires a longer loop.

Other Supersecondary Structures

We have made similar analyses to the other supersecondary structures, the βαβ-unit (Edwards *et al.*, 1987) and the αα-motif (Thornton *et al.*, 1988). The critical determinants of loop conformations seem to be (i) length, (ii) glycines which adopt positive φ values, and (iii) interactions between the loop and hydrophobic or hydrophilic 'anchor' residues, which often form part of the framework, but can determine the loop conformation.

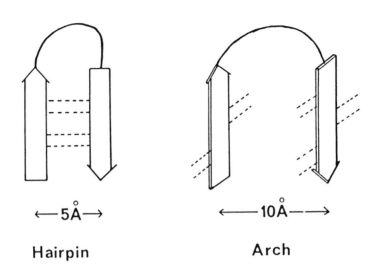

FIGURE 7. A schematic diagram of a β-arch.

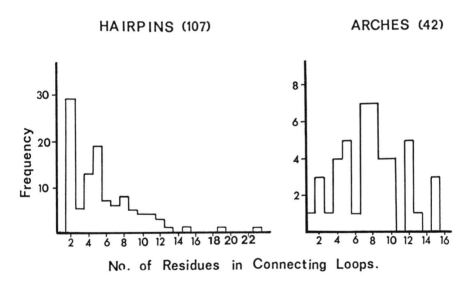

FIGURE 8. Length of loops in β-arches, compared to β-hairpins.

154

Conclusion

Our aim is to develop a hierarchy of sequence patterns stretching from the basic secondary structures to higher levels like the Greek key, the β-meander and the Rossmann fold. We will study these structures in detail to see if characteristics sequence patterns could be used to recognise a specific feature. To do this we have to categorise all of our structures and this is being done as part of our database project.

References

Akrigg, D., Bleasby, A. J., Dix, N. I. M., Findlay, J. B. C., North, A. C. T., Parry-Smith, D., Wootton, J. C., Blundell, T. L., Gardner, S. P., Hayes, F. R. F., Islam, S. A., Sternberg, M. J. E., Thornton, J., Tickle, I. J., and Murray-Rust, P. M. (1988). *Nature*, 335:745–746.

Baker, E. N. (1980). *J. Mol. Biol.*, 141:441–484.

Bernstein, F. C., Koetzle, T. F., Williams, G. J. B., Meyer, E. F., Brice, M. D., Rodgers, J. R., Kennard, O., Shimanouchi, T., and Tasumi, M. (1977). *J. Mol. Biol.*, 122:535–542.

Bleasby, A. and Wootton, J. (1990). *Protein Eng.*, 3:153–159.

Chou, P. M. and Fasman, G. D. (1977). *J. Mol. Biol.*, 115:135–175.

Cohen, G. H., Silverton, E. W., and Davies, D. R. (1981). *J. Mol. Biol.*, 148:449–479.

Edwards, M. S., Sternberg, M. J. E., and Thornton, J. M. (1987). *Protein Eng.*, 1:173–181.

Fujinga, M., Delbaere, L. T. J., Brayer, G. D., and James, M. N. G. (1985). *J. Mol. Biol.*, 183:479–502.

Islam, S. A. and Sternberg, M. J. E. (1989). *Protein Eng.*, 2:431–442.

James, M. N. G., Sielecki, A. R., Brayer, G. D., Delbaere, L. T. J., and Bauer, C. A. (1980). *J. Mol. Biol.*, 144:43–88.

Kabsch, W. and Sander, C. (1983). *Biopolymers*, 22:2257–2637.

Kamphuis, I. G., Kalk, K. H., Swarte, M. B. A., and Drenth, J. (1984). *J. Mol. Biol.*, 179:233–256.

McGregor, M. J., Islam, S. A., and Sternberg, M. J. E. (1987). *J. Mol. Biol.*, 198:295–310.

Milner-White, E. J. and Poet, R. (1986). *Trends Biochem. Sci.*, 12:189–192.

Reid, L. S. and Thornton, J. M. (1989). *Proteins*, 5:170–182.

Sibanda, B. L., Blundell, T. L., and Thornton, J. M. (1989). *J. Mol. Biol.*, 206:759–777.

Sibanda, B. L. and Thornton, J. M. (1985). *Nature*, 316:170–174.

Singh, J. and Thornton, J. M. (1990). *J. Mol. Biol.*, 211:595–615.

Taylor, W. R. and Thornton, J. M. (1984). *J. Mol. Biol.*, 173:487–514.

Thornton, J. M. Sibanda, B. L., Edwards, M. S., and Barlow, D. J. (1988). *Bioessays*, 8:63–69.

Thornton, J. M. and Gardner, S. P. (1989). *TIBS*, 14:300–304.

Wilmot, C. M. and Thornton, J. M. (1988). *J.Mol.Biol.*, 202:221–232.

Wilmot, C. M. and Thornton, J. M. (1990). *Protein Eng.*, 3:479–493.

Discussion

Q: Janet, what sort of side chain interactions do you see?

A: There are only 400 different types of side chain interaction (20 × 20). We have looked at all the known structures, (high resolution) and extracted all pairs of interacting side chains, such as aromatic-aromatic. This work is now published and I refer you to Singh and Thornton (1990). For example, we take all the pairs of interacting arginine-carboxylates and superpose all the arginines onto a reference ARG, thus generating the distribution of the paired residue round that arginine (see Fig. 9). We have now created the database of all 400 PAIR distributions. We also generate the random distributions expected on the basis of chance to assess significance. This sort of information is very difficult to use in ab initio prediction but may be important in modelling side chains and replacements within homologous protein structures. The other aspect we consider is whether these interactions come from helices or strands. For example, we can extract pairs from helical structures only and ask if there is preferred packing between the side chains.

Q: You are dealing with a rather restricted data set in terms of the protein databases. If you are doing developmental work, doing analysis on a restricted data set, I worry that once the data set grows you will need to do all that work again — to update your knowledge.

A: Well, this is one of the absolute requirements that we get the data all into the database as quickly as possible in a completely automated fashion. All the work on the hairpins was done before the database was developed and by hand. The only reason it was possible was by using the assignment program DSSP before it was released (Kabsch and Sander, 1983), so that we could really define accurately or systematically, which is probably more important, exactly where those hairpins were. Once there is an automatic way to extract and classify hairpins, then when we get new data, in a new release of the Brookhaven Protein Data Bank, we will run through the new coordinate files, pull out all derived information, and update the tables. That is why it is so crucial to get the data and extraction procedures set up in an automated way.

Q: Coming back to your side chain interactions. How have you faced the problem at least in most cases that the side chains adopt different conformations, with different X values?

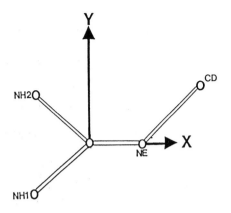

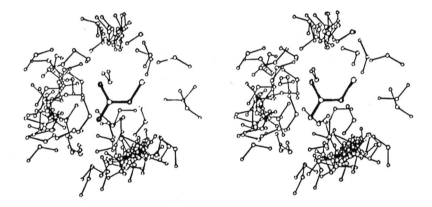

FIGURE 9. The distribution of carboxylate groups around arginine-carboxylate pairs were extracted from high resolution protein structures, and the arginines were all superposed (see text). The resulting distribution shows the preference of the carboxylate to form a pair of hydrogen bonds to the nitrogens of arginine. (see Singh and Thornton, 1990).

A: We use the rigid end groups and those are often the groups which dominate the interaction. But the database is not yet large enough to subdivide the interaction pair data by conformation. Lorne Reid and I attempted to model build side chains given only backbone Cα coordinates (Reid and Thornton, 1989). The idea was to extract from the database all pairs which satisfied a given Cα...Cα constraint and Cα...Cβ directionality. Usually we extracted only three or four residue pairs, which is just not large enough to obtain any statistical significance.

Q: How often are your rules for β-hairpin insertions/deletions applicable?

A: In modelling homologous structures the class 'rules' must always be obeyed (a hairpin by definition can only belong to Class 2, 3, 4 or 5) i.e. if the insertion occurs in the loop the change of class is automatically correct. The rules for predicting the conformation of the loop are usually correct but we have yet to derive percentages. We have recently found some insertions which occur within the strands creating β-bulges rather than in the loops. For the shorter loops a high percentage adopt one of the family structures but for the longer loops no single structures dominate and so no information is available to help predict the conformation of the loop. Other techniques must be used here.

Q: Returning to the side chain database. Do you store the distributions, or can you derive them when you want them?

A: At the moment, we store the contacts, so that we know which residues are interacting. So we can just pull out this information and generate the display from the coordinates.

Q: Main chain interactions are obviously important. Have you considered these?

A: Yes, we have the peptide-pair distributions with all the 20 side chains.

A Review of Methods for Protein Structure Comparison

Christine A. Orengo

Laboratory of Mathematical Biology
National Instistute for Medical Research
The Ridgeway, Mill Hill
London NW7 1AA
U.K.

c_orengo@uk.ac.mrc.nimr

Introduction

As increasing numbers of structures are determined and compared, it has become apparent that there are extensive similarities in the chain fold (or topology) of many proteins. In recent years techniques have been developed which attempt to predict, or model, the structure of a protein from the amino acid sequence by utilising these common structural elements. In addition to identifying recurring topologies, structure comparisons have also revealed smaller structural fragments or motifs accross a wide range of proteins; for example, the $\beta\alpha\beta$ and the β-hairpin motifs (see Thornton *et al.*, in this Vol.). Characterising these motifs in terms of geometry, distance constraints and sequence composition can enhance protein structure prediction techniques by incorporating longer range interactions between secondary structures.

Structure comparison methods can be divided into a number of approaches. Some employ a variety of techniques and are not easily categorised under one heading. They are classified below according to their most novel aspect and for each approach, developments and innovations are described in chronological order:

- methods compairing inter-residue distances in the two proteins methods comparing the geometry of fragments along the polypeptide chain

- methods superposing the two structures or fragments of the two structures

- methods comparing the relative dimensions and orientations of secondary structures in two proteins

- methods using dynamic programming techniques to align inter-residue distances, structural fragments or other properties between the two proteins

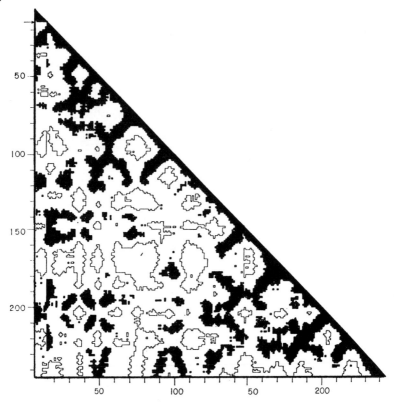

FIGURE 1. The distance map for α-chymotrypsin. Black region: $d_{ij} \leq 15\,\text{Å}$; closed region: $d_{ij} > 35\,\text{Å}$, (Nishikawa and Ooi, 1974).

Methods Based on Comparison of Inter-Residue Distances

Distance Plots

In 1970, Phillips devised a useful qualitative method for recognising regions of secondary and supersecondary structures in a protein. A two dimensional matrix, known as a distance plot was used to display all the inter-residue distances within the protein. Both axes of the plot correspond to the amino acid residues of the protein and individual cells are shaded depending on the distances between respective pairs of residues. Contour lines can be drawn at fixed inter-residue distances and the patterns which arise are characteristic for different types of secondary and supersecondary structure (Fig. 1).

An analysis and description of these patterns, associated with different levels in the hierarchy of protein folding, has been provided by Kuntz *et al.* (1975). For example in secondary structures (e.g. α-helix or extended β-strands) residues are close together giving density near the diagonal, while packed tertiary structures are revealed at greater distances from the diagonal. Patterns due to turns and loops appear first and subsequently patterns due to interacting secondary structures. Two parallel secondary structures, such as β-strands produce a line parallel but at some distance from the diagonal whilst antiparallel segments would give rise to a series of contour lines perpendicular to the diagonal. Kuntz *et al.* (1975) describe a method for recognizing supersecondary units from the square or trapezoidal patterns which they generate. These are referred to as contour cells and contain between 30 and 50 residues.

Because the distance matrix is invariant to rotation and translation of the corresponding molecule, it is useful for comparing protein structures and similar local folds and subdomains can be recognized by matching patterns in the two distance plots. For more quantitative comparisons, Nishikawa and Ooi (1972) generated difference distance plots which display differences in distances between corresponding points in two distance plots.

Method of Padlan and Davies

Padlan and Davies (1975) originally used difference distance plots to compare the constant and variable immunoglobulin domains (see Fig. 2). A problem can arise where structures being compared are not equal in length, as corresponding patterns occur in different positions in the two distance plots. One method of dealing with this for homologous proteins, is to simply exclude any inserted residues between the two proteins. Where the homology is less clear, they use a rather rough method for normalising the lengths of the two proteins. Additional equispaced points between equivalent marker positions in the two proteins are constructed, such that both polypeptide chains have equal numbers of points between any two markers. These marker positions must be clearly structurally homologous and generally lie in the interior of the domain. As the structures become more remote it becomes harder to accurately determine regions of structural homology. Another disadvantage is that by generating average positions in regions with insertions and deletions, it is possible to obscure any similarity which may be present.

Sippl's Method

A more recent development of the method due to Sippl *et al.* (1982) and referred to as the D_k procedure, compares subsets of the distance matrices of the two proteins

a   b

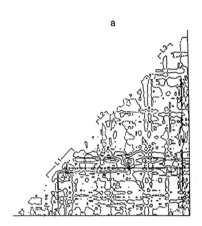
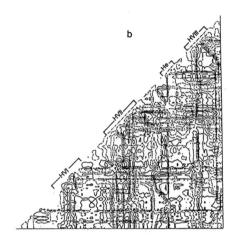

FIGURE 2. Difference distance plots for the comparisons of (a) variable light domain of McPc603, with the variable light domain of REI and (b) the variable light domain of McPC603 withe the variable heavy domain. The contours are at 2.5 Å intervals; positive contours are solid, zero and negative are dotted. HVI, HVII and HVIII encompass the regions where the light chain, heavy chain or both residues are hypervariable. Reproduced by kind permission of the Proceedings for the National Academy of Sciences.

being compared. These subsets are generated from different diagonals of the distance matrices and contain elements $d_{i,i+k}$ referring to the distance between the ith and $i + k$th residue where k represents the order of the diagonal relative to the main diagonal.

$$d_{i,i+k} = |(x_i - x_{i+k})^2 + (y_i - y_{i+k})^2 + (z_i - z_{i+k})^2|^{\frac{1}{2}}$$

The diagonal deviation of order k between two proteins A and B (distances expressed as d and $\overline{d}$ below) can therefore be expressed as:

$$D_k(\overline{S}, S) = \left[\frac{1}{N-k} \sum_{i=1}^{N-k} \left(\overline{d}_{i,i+k} - d_{i,i+k} \right)^2 \right]^{\frac{1}{2}}$$

where $\overline{S}$ and S are the structures being compared of equal length N. To compare two molecules a set of D_k values for $k = 1 \ldots, N - 1$ is defined as the D_k spectrum of the deviations of the two stuctures.

Unlike the rigid body superposition methods discussed in more detail below, the D_k method can give information on internal rotations occurring between two otherwise similar molecules. If you compare low order diagonals between two distance plots (that is, diagonals which display distances between residues close in the sequence), and plot differences against sequence position, then peaks will occur in the region of the internal rotation. In other regions the similarity in the diagonals is maintained.

Generally, difference plots can indicate whether similarities are distributed uniformly along the sequence or occur at particular positions. Another useful property is that different diagonals reveal similarities at different levels of structural organisation. For instance, for secondary structures, diagonals of order up to $k = 10$ should be used, whilst for domains $10 \leq k \leq 25$ diagonals will give medium and long range structural information. Only a few D_k values should be necessary to give a good analysis of structural similarity between two proteins.

Methods Comparing the Geometry of Fragments Along the Polypeptide Chain

Rackovsky's Method

Rackovsky and Scheraga (1978, 1980, 1984), devised a representation of proteins based on differental geometry. This treats the polypeptide backbone as a simple curve and takes measurements of local geometry at regular intervals. Such a representation allows structural comparisons to be performed in a simple manner. The method uses fragments of four α-carbons, and is therefore designed to operate on a length scale intermediate between distance plots, which can incorporate long range distances and methods based on individual residue comparisons. A four α-carbon unit is chosen as the smallest length of polypeptide chain which can fold.

Two parameters are calculated at each $C\alpha$-residue, defining the curvature (κ) and torsion (τ) of the chain over the next successive four-$C\alpha$ fragment (see Fig. 3). The authors have characterised the regions in (κ, τ) space occupied by various types of secondary structure. To compare two protein structures, differences between their curvatures and torsions are plotted as a function of position along the chain. Thus any change in local geometry appears as a peak along the sequence. The method is sensitive to the presence of an internal rotation between similar structures as plots of κ, τ will show complete identity except in the region of the rotation. However, comparisons between remote structures will be affected by insertions/deletions which will obscure structural similarities occurring at different positions along the sequence.

A recent development of the approach (Rackovsky, 1990) uses other geometric measurements to further quantify successive four $C\alpha$ fragments (see Fig. 4). Fragment

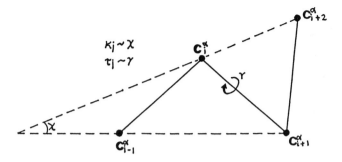

FIGURE 3. Definition of κ_i and τ_i. The curvature is proportional to the angle between the $C\alpha_{i-1} - C\alpha_{i+1}$ and $C\alpha_i - C\alpha_{i+2}$ vectors. The torsion is proportional to the dihedral angle γ for rotation about the $C\alpha_i - C\alpha_{i+1}$ virtual bond. In general, the four-$C\alpha$ unit is not planar.

types are then characterised depending on the ranges to which their distances and angles belong, so that a fingerprint can be constructed for each protein in the database described by the distribution of fragment types within the structure.

Proteins are compared by using a metric function to determine the distance between their fingerprints (this effectively calculates a root mean square deviation between the distributions of fragments in the two proteins). The authors have compared all the structures in the databank and subsequently applied graph theory and cluster analysis to identify 'galaxies' of similar proteins. The positions and relationships between these clusters indicates that proteins form a continuum extending from mainly extended to almost entirely helical structures, though the density of structures is not uniform accross this continuum. The largest clusters are associated with the nucleotide binding proteins (alternating β/α class) and the all-α helical proteins.

However, the analysis has not yet been extended beyond fragments of four α-carbon length, which would be of interest regarding longer range interactions. Currently, the method does little more than might be obtained using secondary structure compositions and gives no information about the relative positions of similarities in any two proteins.

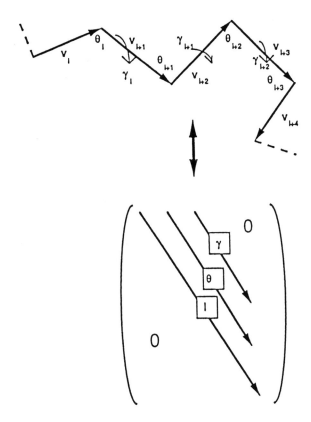

FIGURE 4. The relationship between the structure of a backbone fragment and the generalised bond matrix that represents it. An N-vertex fragment is represented by a $(N-1) \times (N-1)$ matrix.

Levine's Method

An alternative method based on comparing torsional angles along fragments of polypeptide chain, was developed by Levine et al. (1984) as a screening method for ranking structural similarities between pairs of proteins. Differences between any two proteins is represented by a two dimensional matrix whose ijth element is given by the following expression:

$$\Delta_{ij} = (\Delta\phi + \Delta\psi)_{ij} = |\phi_i - \phi_j| + |\psi_i - \psi_j|$$

Similarity in dihedral angles ($\phi\psi$) will result in a run of similarly shaded cells in the matrix. As for distance plots, insertions in one of the structures will displace this run sideways. Once the matrix has been filled with Δ_{ij} values, it can be analysed in a number of ways to obtain an overall similarity score between the two proteins. These include a fast method of searching for the best path through the matrix, applying the techniques of list processing. A second method uses statistics to obtain a measure of similarity based on the distribution of Δ_{ij} values along all the diagonals. As the authors point out, the statistical nature of this method may lead to the significance of some comparisons being missed. A further problem is that similarities in secondary structure will be identified in the matrix regardless of topological equivalence and this will complicate the measurement of structural equivalence.

Karpen Variation

A similar technique based on comparing torsional angles along the polypeptide chain, has been devised by Karpen et al. (1989). The algorithm computes Δ_t, the root mean square difference in ϕ and ψ torsional angles over a small number of amino acids (usually 3–5). To compare two proteins, all n-length substructures from one protein are compared with all possible n-length substructures from the other. Values of n will depend on the application. For example to find small local regions of structural similarities such as β-turns, substructures of less than 10 are used.

Regions of similarity are again visualised using a two dimensional matrix, with axes corresponding to the amino acid residues of the proteins. Each element (i, j) is shaded according to the value of Δ_t for the comparison of substructures starting at residue i in protein 1 and j in protein 2. The lower the value of Δ_t, the darker the square. Similar secondary structures appear as low-valued rectangles. Several substructure sizes were tested, large values tend to obscure similarities which do not extend over the whole substructure, whilst small fragments will generate too many trivial similarities. A substructure length of 3–5 is recommended for detecting regions of similarity.

To establish probabilities and values of Δ_t reflecting structural similarity, the distribution of Δ_t was obtained for a large number of substructure comparisons, involving proteins from each of the protein classes. Threshold cutoff values in the range of $25°$ to $35°$ are optimal for differentiating between similar and non-similar substructures. Because the method is fast, it can be used as an initial scan across the databank. As with distance plot methods, topological similarities between remote structures will be hard to identify, but the method can successfully locate backbone conformational differences between similar structures and is therefore useful for identifying possible active sites and ligand binding regions.

Method of Nussinov and Wolfson

A potentially fast algorithm for structure comparison also using protein geometry has been developed by Nussinov and Wolfson (1991). This utilises object recognition techniques, originally investigated for computer vision. A protein molecule is defined as a (geometric) rigid object consisting of a set of points corresponding to atoms. Given a new protein and a database of known proteins, the algorithm can find any protein having large identical substructures with the observed protein.

For each object, coordinate frames or reference sets are generated from all possible triplets of ordered non-collinear atoms in the molecule. Atoms in the molecule are then represented by their coordinates in these reference sets. The points in each molecule (or object), together with associated labelling and geometric constraints are memorised in a vast table indexed by the integer values of the x, y, z coordinates. Once compiled, the table can be used for any comparison and any new proteins can be processed without recomputing the table. To match a new protein against the database, successive reference sets are generated for the new protein, coordinates computed and then tested against the stored points in the hash table. Any reference set in the database having a large number of matching points, is taken as corresponding to that chosen for the new protein and can be used to calculate the transformation of the new protein to the relevant database protein, giving a superposition of the two structures or substructures.

The method has not been coded in a computer program and so remains untested. However, it potentially has some problems with the look-up table. Firstly, this may simply be too large to store and secondly, a more fundamental problem is that two atoms are assumed to be equivalent only if they have the same index, which means they must correspond within 1Å and except for closely related structures this may not be so.

Methods Which Superpose Protein Structures or Fragments of Structure

The first application of a simple superposition method was by Cullis *et al.* in 1962 who compared α and β horse haemoglobin and sperm whale myoglobin, by orientiation of low resolution views of their structures. Subsequently, a number of other groups developed techniques based on rotating the structures relative to each other in order to minimise the sum of the squares of the distances between equivalent backbone atoms.

A more quantitative approach appeared in 1973 when Rao and Rossman described a method based on finding a rotation matrix C and a translation vector d which superposes two similar structures in such a way as to minimise the sum of the squared distances between all equivalent atoms (N).

$$X'_{II.i} = CX_{I.i} + d$$

If $X_{I.i}$ and $X_{II.i}$ are the position vectors of the ith equivalent atoms in molecules I and II relative to two arbitrarily chosen axes, the deviation between the structures is then minimised:

$$S = \sum_{i=1}^{N} \left(X_{II.i} - X'_{II.i} \right)^2$$

Determination of C and d is a three step process. In the first step, parameters are evaluated by a least squares procedure but as this can give rise to anisotropic expansion of one molecule relative to the other, the next step regains rigidity by examining the rotation matrix and deriving an approximate set of rotation (eulerian) angles. In the final step, these angles are used to determine the true minimum for S above by another non-linear least squares minimisation. If there are substantial differences or conformational changes between the structures then the last step has to be repeated leaving out atoms which are separated by more than some threshold distance, until only equivalent atoms remain. The method can be hard to apply when there is no knowledge of equivalent atoms between the structures.

Rossman and Argos (1975, 1976) use an iterative method to determine the set of N residues which have the highest probability of being equivalent. Probabilities, related to spatial similarity and orientation are determined for each residue pair between the two proteins and those residue pairs having the highest probabilities and obeying the rule of sequential selection are chosen.

$$P_{ij} \propto \exp\left(\frac{-d_{ij}^2}{2E_1^2}\right) \exp\left(\frac{-S_{ij}^2}{2E_2^2}\right)$$

where

$$d_{ij} = X_{I.i} - X_{II.j},$$

$$S_{ij}^2 = (d_{ij} - d_{i+1,j+1})^2 + (d_{ij} - d_{i-1,j-1})^2,$$

E_1 and E_2 are empirical weighing factors and S_{ij}^2 measures the scatter of vectors between successive superposed residue pairs. After each superposition, the set of equivalent residues is determined and the molecules resuperposed using this set until there is no change in equivalent residues.

An advantage of the method is that it can accommodate limited insertions and deletions as long as sets of equivalent atoms can be identified obeying above the progression rule. A small disadvantage is that any transposed supersecondary structures or domains may not be recognised between proteins. Also the fact that arbitrary values are set for various parameters used in the search for initial equivalences can be a problem and Zuker and Somorjai (1989) suggest that the method often does not converge reliably and can lead to inconsistencies. Furthermore as the method only generates a single score, there is no information quantifying degrees of similarity in different regions of the proteins.

Method of Remington and Matthews

A different approach was developed by Remington and Matthews (1978, 1980) based on Fitch's method of comparing amino acid sequences (see Fig. 5). Two proteins are compared by optimising agreement between all possible segments of a chosen length from one protein with all possible segments of the same length from the other protein.

Application of the method is straightforward. A suitable probe length, say 40 residues is chosen, and then each 40 length segment of one protein is compared with each 40 length segment of the other. Segments are rotated and translated so as to minimise the root mean square (RMS) deviation, $R_{C\alpha}^{ij}$, between respective α-carbons (see Equation 1 below). If the segment length is too small, the average structure agreement is found to vary with protein class. A segment length of 40–80 residues is recommended for

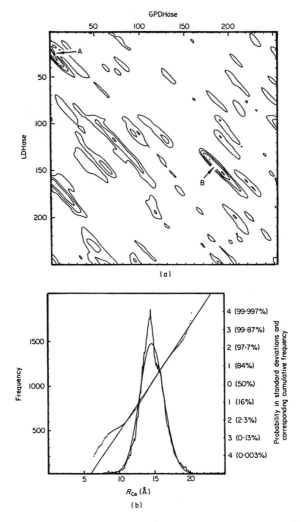

FIGURE 5. (a) Structure comparison map for lactate dehydrogenase with glyceralde-hyde-3-phosphate dehydrogenase with a probe length of 80 residues. Successive contour levels indicate root mean square differences between fragments ($R_{C\alpha}$ of 13.0 Å, 11.3 Å...) and are at intervals of one standard below the mean value of 14.73 Å. Peak A indicates the alignment of the respective nucleotide binding folds and peak B the similarity in the structures of the catalytic domains – deviation (1.72 Å) below the mean value of 14.73 Å. (b) Frequency distribution of $R_{C\alpha}$ for the comparison of lactate dehydrogenase with glyceraldehyde-3-phosphate dehydro-genase with a probe length of 80 residues. The best fit Gaussian distribution is superimposed. The figure also includes the same data plotted as a cumulative distribution. Values of $R_{C\alpha}$ are grouped in increments of 0.1 Å unit. The ordinate on the right gives the probability in units of 1s and the corresponding cumulative frequency of $R_{C\alpha}$, expressed as a percentage. Reproduced by kind permission of the Journal of Molecular Biology.

identifying topological similarities between proteins.

$$R_{C\alpha}^{ij} = \frac{1}{L} \left[\sum_{k=0}^{L} \left(X_{i+k} - A(\alpha, \beta, \gamma) X_{j+k}' \right)^2 \right]^{\frac{1}{2}} \tag{1}$$

where X_i and X_j' are coordinate vectors of the ith $C\alpha$ atom and the jth $C\alpha$ atom of the two proteins relative to the centre of mass of the L atoms being compared. A is the rotation matrix for rotation of X' about the x, y, z axes respectively.

Application of McLachlan's fast matrix minimisation method (1982) allows all possible comparisons to be performed in a reasonable amount of time and one of the advantages of the method is that it produces a large sample of values against which a good result can be assessed.

The method is analogous to the dot matrix plot used in sequence comparison and is therefore useful for identifying structural repeats. Similar segments in separate proteins can be located provided they are not interrupted by large insertions and deletions.

Methods Based On the Geometry of Secondary Structures

A number of different methods exist for matching secondary structures between proteins. Most employ one of the techniques described earlier, such as distance plot analysis or superposition.

Richards and Kundrot (1988) adopt an approach based on distance plot analysis. Secondary structures are first determined by comparing distances in the real structure with those in idealised models of secondary structure. In the second step, the identified helices and strands are represented as straight line segments and for each pair of secondary structures the geometric relations between their line segments are defined by six parameters.

Elements of secondary structure and the relations between them are displayed in a two dimensional character matrix (see Fig. 6), the axes of which represent the amino acids of the protein. Interacting secondary structures are identified by off-diagonal submatrices or 'boxes' which also display the angles between the axes of any two elements. To search for a structural motif within a protein, a mask distance matrix of the motif is generated and compared to that of the protein, scoring the fit between the query and protein matrix at each position in the latter. Overall fit is measured by calculating the root mean square differences between the two matrices. Obviously sensitivity is affected by the presence of insertions/deletions.

172

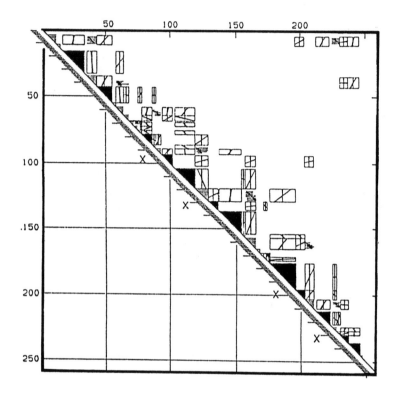

FIGURE 6. Character matrix representation of triose phosphate isomerase, PDB file – 1TIM showing secondary structure, H bonded strand pairs and interaxial angles for helix-helix, helix-strand and strand-helix interactions. The axes display the residue numbers.

In a development of the method, metamatrices, are constructed, with the axes representing secondary structures rather than residues. Off diagonal cells describe relations between secondary structure segments and can contain one to six parameters depending on the type of metamatrix. Comparisons of metamatrices to identify substructures should be faster and less affected by insertion/deletions. However, the authors have not yet described the full implementation of this approach.

Abagyan and Maiorov (1988) use a simplified quantitative representation of the polypeptide backbone to search for similar fragments of structure in the protein data base. Sets of vectors are used to represent the folding pathway of the protein chain. Each regular segment (α-helix or β-strand) is represented by a vector along its axis from the N to the C terminal. Vectors representing non-regular segments are constructed between the endpoints of the regular segments (see Fig. 7). Spatial organisation of the vectors is described by the lengths of the vectors and the planar and dihedral angles between them. Superposition of the vector representations of two structures can then be performed using McLachlan's algorithm (1982).

Although there may be problems in allocating residues to regular or non-regular secondary structure elements, this method will be better able to accommodate insertions and deletions as these tend to occur in loops between the secondary structures. An additional problem though, is the curvature of secondary structure elements. Although helices can generally be represented as straight lines, β-strands often have to be divided into two or three vectors as they can be more substantially curved.

The main advantage is that the small number of parameters and the use of length and angles instead of coordinates enables fast automatic recognition of topologically similar fragments in the structure database. So far, it has only been tested on small toxin structures and has yet to be used to search for structural similarities in larger proteins.

Murthy's Method

A related approach which also incorporates the techniques of dynamic programming, has been developed by Murthy (1984). In the first step a similar vector representation of helices and strands is generated. One molecule is then rotated relative to the other and the probability that two elements in the structures are equivalent is given by:

$$P_{ij} = w_{ij} \cos \theta_{ij}^n$$

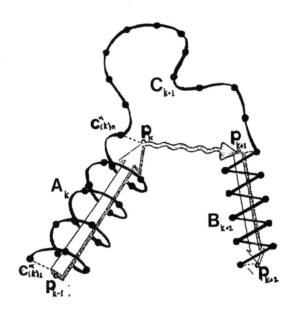

FIGURE 7. Representation of a protein polypeptide chain as that composed of regular secondary structure and coil segments. The regular segment, for example α-helix (A) and β-strand (B), is represented by a vector directed from the N- to C-terminal residue of the segment and situated on the axis. The points of the start (P_{k-1}) and the end (P_k) of the vector are determined by projecting the first and last C_α atoms onto the segment axis. The vector corresponding to the non-regular segment connects the end of the preceding regular segment (P_k) with the start of the next regular one (P_{k+1}). Reproduced by kind permission of the Journal of Biomolecular Structure and Dynamics.

where θ_{ij} is the angle between the ith secondary structure element of molecule 1 and the jth secondary structure element of molecule 2 and n is chosen to be between 5 and 10. The weight is zero if the elements are different types and can be adjusted depending on the number of atoms in the element and the distances between the atoms. At the correct orientation P_{ij} values between structurally equivalent elements will be high.

Probability values (P_{ij}) for a given orientation are placed in a two dimensional matrix which is evaluated using dynamic programming (see below). The score returned from this is then modified by a term dependent on how well the vectors of equivalent secondary structures in the molecules have been superposed.

$$\sum_{ii} \exp\left(\frac{-|\Delta d_{ij}|^2}{2E^2}\right)$$

where d_{ij} is the difference between vectors from the ith to the jth equivalent secondary structures in the two molecules and E is the RMS value of the scatter between vectors.

Finally, the modified scores are plotted as a funtion of the 3 eulerian angles. Apart from identifying the orientation giving the best superposition of the structures, θ and d_{ij} of the peak also give information on changes in packing of secondary structure elements between the molecules. This type of information is not easily obtained from the Rossman and Argos (1975, 1976, 1977) type superposition methods which return a single value reflecting the overall similarity of the two structures.

As with the method of Abagyan and Mairov, the method is very fast because there are relatively few secondary structures as opposed to the large number of residues comparisons which have to be evaluated in alternative superposition methods.

Graph Method of Mitchell *et al.*

With the aim of producing a fast method for databank searches, Mitchell *et al.* (1989) used graph theory to compare secondary structure geometry between two proteins. Firstly, the linear represention of each secondary structure is determined by calculating the centre of mass from coordinates of the constituent α-carbon atoms. An 'ideal' secondary structure is fitted and the endpoints of the axis determined. Curved structures are approximated by more than one line.

Angles between pairs of axes, together with distances between midpoints and closest approach distances are evaluated and stored. Each protein or motif can then be represented by a graph, the nodes of which correspond to the linear secondary structures, and the edges to the inter-line angles and distances. It is possible to check whether a query graph is contained within one or more of the stored graphs of databank structures by application of subgraph isomorphism algorithms (Ullman, 1976).

Trial searches were performed for detection of TIM type β-barrels, nucleotide binding folds and calcium binding folds. Those that failed were generally distorted in some way, for example with strands more twisted than in the query structure or missing one or more secondary structures contained in the query structure. This suggests that although the approach is very fast (a search for a motif of about 15 secondary structures taking roughly 30 minutes) and reliably detects structures close to that of the query, there may be some problems recognising structures or fragments with greater deviations and perhaps with absent or additional secondary stuctures.

Methods Incorporating Dynamic Programming Techniques

This section describes methods which exploit the techniques of dynamic programming. These were introduced by Needleman and Wunsch (1970) for the purposes of aligning amino acid sequences and are extremely well suited to accommodate the presence of insertions and deletions.

In the first instance, a two-dimensional matrix representing the similarity of the two proteins is scored by comparing identities and/or physico-chemical properties of residues. Scores are then accumulated across the matrix by adding the score from each cell to all the cells in the row and column to the left and above the current cell. Equivalent residues in the two proteins are identified by the highest scoring pathway through the matrix. This is found by starting at the N terminus and tracing a path through the matrix, taking the highest scoring cell at each step. The path is not allowed to double back across any row or column. Insertions and deletions can be accommodated by allowing gaps in either of the sequences and result in the path adopting a horizontal or vertical course in the region of the gap. An adjustable gap penalty will restrict the number of gaps depending on the expected similarity of the proteins (see Barton in this Vol. for a fuller explanation).

Method of Zuker and Somorjai

Zuker and Somorjai (1989) combined both dynamic programming and superposition techniques. Their algorithm finds a set of non-overlapping fragments from one protein (A) and a set of matching fragments from a second protein B, which can be optimally superimposed on the set from A. Corresponding fragments must be of equal sizes and contain at least three residues.

The k fragments are superposed by rotations and translations such that the following quantity, referred to as the minimum distance between proteins A and B is minimised:

$$D(A, B) =$$

$$\sum_{h=1}^{k} \sum_{i=0}^{m_{2h}-m_{2h-1}-1} \| A\left(m_{2h-1}+i\right) - \Re_h\left[B\left(n_{2h-1}+i\right)-T_h\right] \|^2 +$$

$$\alpha(k-1) + \beta\left\{ m+n-2\sum_{h=1}^{k}\left(m_{2h}-m_{2h-1}+1\right) \right\} \tag{2}$$

where $\Re_h$ is the rotation matrix for the hth fragment and m and n represent residues in proteins A (length M) and B (length N) respectively. The first part of the equation is obtained from superposition of the k fragments. The inner sum determines the optimum size and superposition of the hth fragment. The second part is the penalty for forming a break between two fragments and is used to inhibit the solution from degenerating to the trivial case of superposing fragments of length three. Finally, the third part is the penalty charged for every residue left out of the alignment. Scores proportional to the difference (D) between any two positions in the proteins are stored in a two dimensional similarity matrix.

Since a large number of superpositions are required to determine the best sized fragment at each position in this matrix ($M \times N$), Zuker and Somorjai developed a fast superposition method. The rotation matrix is transformed into a more symmetric form by the use of quaternions. The solution to the superposition problem is then obtained using the maximum eigenvalue method, which is significantly faster than the non-linear iterative method required using eulerian angles.

The alignment of the two structures is solved recursively by building up from the amino terminals. $V(i)$ is defined as the minimum value of Equation 2 above for aligning coordinates $A(1), A(2)\ldots, A(i)$ with $B(1), B(2)\ldots, B(j)$. There are 4 possible cases for each value of i, j depending on whether residues $A(i)$ and $B(j)$ are aligned or not and depending on whether any fragments of structure have been superposed within these segments. Computation of k and the fragments giving the best alignment is achieved by a traceback through the two-dimensional (V) matrix to find which of the four conditions is best satisfied at each step.

All pairwise comparisons, calculated the first time the structures are aligned, are saved. This means different values of α and β can be tested in order to fine tune the alignment. Furthermore, the method can also be used as a simple database searching algorithm by using a template taken from one protein structure and sliding it along the backbone of the second structure, performing a superposition at each stage.

One problem of the method is that for more remote structures (e.g. hen against T4 lysozymes), some values of the gap penalties cause a large number of very small fragments to be equivalenced, so that little information regarding topological similarity is generated. A further disadvantage is that it is not possible to take account of insertions/deletions (indels) within fragments. The only way to deal with these is to create a separate fragment. Zuker and Somorjai (1989) discuss the possibility of distinguishing between sequence indels within fragments and real three dimensional breaks and penalising them separately, but have not explored these options yet.

Săli and Blundell's Method

The hierarchic organisation of protein structure is exploited by the method of Săli and Blundell (1990) which uses a multilevel representation. The protein is described as an indexed string of elements that can exist at several hierarchical levels: residue, secondary structure, supersecondary structure, motif, domain or globular struture.

Every element was allocated features which relate to the protein fold. These can be individual properties of the element or they can be associated with relationships between elements. For example, an individual property of an element at the residue level could be size or hydrophobicity. Other properties include: accessabilities, dihedral angles, charge and so on. Higher levels of structure such as secondary structure, have properties such as accessability and orientation of the vector through the secondary structure with respect to centre of gravity. Comparison of these properties are stored in an N by M similarity matrix where N and M are the respective lengths of the proteins being compared. Alignment of the proteins is obtained from this matrix using Needleman and Wunsch techniques.

Relationships between elements such as hydrogen bonds and packing are more difficult to compare since these offer a wide choice of potential partners with which to relate. Therefore they cannot be compared by conventional dynamic programming approaches. Instead equivalences are determined using simulated annealing which effectively searches through possible partners to find the best pairs. This process must be tried several times, as the technique does not necessarily find the global minimum. However, the authors suggest this should not affect the result, as any score close to the minimum will help in finding the correct alignment.

After comparing both properties (p) and relationships (r) at each level, the weighted sum of all the comparisons between two elements over all levels (l) in the hierarchy is expressed by:

$$W_{ij} = \sum_l \left(\sum_p \rho^p \, {}^n w_{ij}^p + \sum_r \rho^r \, {}^n w_{ij}^r \right)$$

where ${}^n w_{ij}^f$ is the normalised difference in feature f between residues i and j from the first and second proteins. The scaling factor ρ^f determines the relative importance of any feature f. These weighted sums are placed in the weight matrix W, using a function which counts how many times elements i and j are matched. The matrix is then evaluated using the Needleman and Wunsch algorithm.

Although the method uses a very flexible definition of topological equivalence which should help in the alignment of remote structures, this also means it is less automatic in its application, as it is necessary to identify which features to compare and set up corresponding weights. Furthermore the simulated annealing step can be quite time consuming for remote structures.

Method of Taylor and Orengo

Taylor and Orengo (1989a,b) use dynamic programming in an approach which is based on the idea of using distances within proteins to compare structures. For each residue in a protein a local structural environment is defined by the set of vectors from the β-carbon of the residue to the β-carbons of all other residues in the same structure. Residues in two proteins are then matched by comparing their structural environments (see Fig. 8). Because these are defined independently for each residue, in the coordinate frame of the α-carbon, they are rotationally invariant which makes their comparison insensitive to the displacement of substructures.

As with sequence alignment a similarity matrix, known in this method as the *residue* matrix, is used to score the similarity of all residues pairs between the proteins. To compare each residue pair, other similarity matrices, called *distance* matrices are constructed. These represent the comparison of the residue structural environments or vector sets. The *distance* matrices are scored by subtracting interatomic vectors and are evaluated using dynamic programming. The resulting pathways are then accumulated in the residue matrix by summing scores in corresponding cells (see Fig. 9). This has the advantage of reinforcing the alignment of structurally similar regions.

Other aspects of protein structure and composition such as residue accessability, torsional angles or hydrogen bonding patterns can also be included to improve the alignment of remote structures. Scores for matching these properties are added to the

A

B

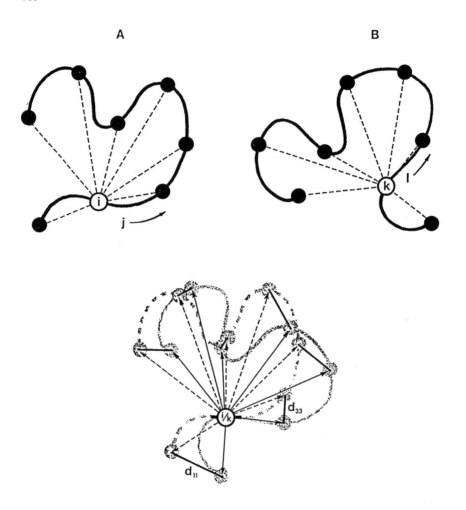

FIGURE 8. Structure comparison by the method of Taylor and Orengo. The two chains A and B are simple two dimensional representations of two similar protein structures. Two positions in these structures, i in A and k in B are compared. In C the structures are aligned on residues i and k and the distances between positions (all j in A and all l in B) are compiled in a matrix. To avoid confusion, only the distances between sequentially equivalent positions are drawn in C (these constitute the diagonal of the matrix). This matrix is then processed by a sequence alignment algorithm and the best correspondance of positions found. The process is repeated for all pairs of positions (all possible locations of i in A and k in B) and the results accumulated into an overall concensus alignment. Reproduced by kind permission of Protein Engineering.

residue or distance matrices depending on whether they reflect individual properties or relationships between residues. Parameters for incorporating different matching options have been optimised so that the method is completely automatic requiring only residue coordinates and residue properties. The comparison of hydrogen bonding patterns is solved by recasting the bond network as a distance matrix. Extended paths across the network of equal length are then matched by comparing an accumulated 'bond energy' gathered over the path.

A more recent and faster version matches only subsets of residues from the two proteins. Comparisons are initially allowed on the basis of similarity in residue accessability and torsional angles and are used to give a preliminary scoring of the residue matrix which identifies the most equivalent residues in the two proteins. Subsequent comparisons using these equivalent residue pairs reduces the amount of noise in the residue score matrix and facilitates the recognition of the correct alignment pathway (see Fig. 10). The set of equivalent residues are iteratively refined by identifying the highest scoring residue pairs after each structural alignment. Up to three passes through the alignment procedure are used (see Fig. 11). Because of the considerable reduction in the number of comparisons an increase in speed of between 50 and 150 fold is achieved allowing comparisons to be made in the order of a few tens of seconds. The ability to include different matching options and the increased accuracy of the alignment obtaining by restricting comparisons to residues in similar structural locations, has improved alignments between remote structures.

Since the method of Taylor and Orengo uses only a sequence alignment algorithm, this can be easily substituted by any of the many alternative approaches to this problem such as the method of Smith and Waterman (for best local alignment) or that of Sellers for multiple sub-alignments. We are currently using the latter to identify similar subunits and structural motifs between proteins in different families. Furthermore, the method generates scores reflecting structural similarity for each aligned residue pair and these should help in the classification of structural constraints for any new motifs discovered.

The direct correspondance of the sequence problem has also allowed the method to be incorporated directly into the fast multiple sequence alignment method of Taylor (1988). Although this is based on pair alignments the final result utilises intermediate consensus sequences. The correspondance of these in structure terms is a bundle of vectors for each position for which an average direction and an error can be calculated and used for comparison.

More recently, the comparison method of Taylor and Orengo (1989a) has been adapted to align proteins by secondary structure matching (Orengo *et al.* (1991)). The substantial increases in speed allow the method to be used as a fast prescan of the database (pairwise alignments between all members of the databank can be generated in under four hours). Subsequently a more detailed and informative comparison of related structures can be performed by residue matching using the rough correspondance generated at the secondary structure level.

Secondary structures are linearly represented by vectors along the central axes and a number of properties determined, including total hydrophobicity and accessability. Also defined are the relationships between secondary structures, expressed by several parameters including tilt/rotation angles, overlap and the vectors/distances between midpoints. Local coordinate frames of reference for determining vectors, are centered on the midpoint and based on the secondary structure axis and the vector to the next successive midpoint along the sequence.

Both individual properties and relations of secondary structures can be compared in a similar manner as for residue matching. Parameters have been optimised using three data sets, a set of globin structures, a set of immunoglobin structures and a set of more remote $\alpha\beta$ nucleotide binding structures. The best alignments were obtained by matching vectors between midpoints and tilt/rotation angles.

As only secondary structure elements are matched, there are fewer comparisons and therefore less time is required than for alignments based on residue matching ($\sim$ 2–5 CPU seconds on a SUN4/280, for large proteins containing more than 20 secondary structures, as opposed to $\sim$ 10–50 CPU seconds on a SUN4/280, using residue matching). This means that the method can be used to compare the entire database of known structures, looking for similar structures or fragments and the large sample of scores generated provides a background assessing the significance of any individual score.

---

[1]The method is used at two levels of comparison. First, to find the best equivalence of distances for the 2 residues being compared, then at a higher level to find the best equivalence of residues within the 2 sequences being compared. (a) the score matrix between two peptides HSERRHVF and GQVGMAC. (b) The score matrix for comparison of all distances centered on residue C in sequence B with all distances centered on residue F in sequence A. The dynamic programming algorithm is used to find the best pathway through this matrix and the values along this path are then accumulated in the corresponding cells of the higher level matrix (a). (c) The lower level process is repeated for residue C (sequence B) and V (sequence A). When all residue pairs have been compared in this way and the values accumulated in matrix (a), the dynamic programming algorithm is then used to find the best pathway through matrix (a).

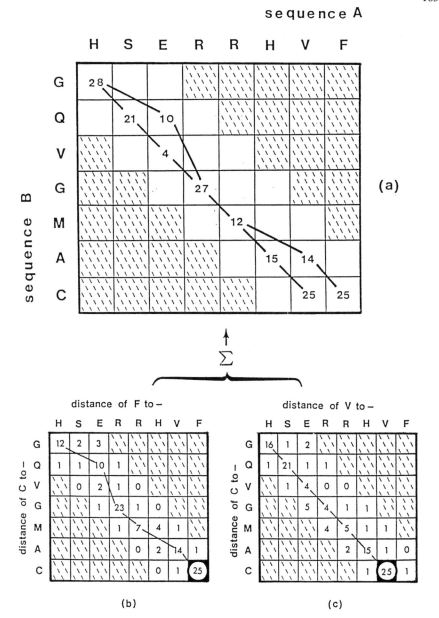

FIGURE 9. Application of the dynamic programming method to structure alignment in the method of Taylor and Orengo[1].

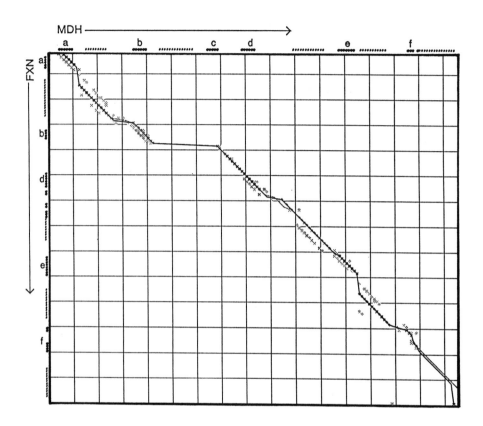

FIGURE 10. Alignment of malate dehydrogenase with flavodoxin by the three pass method which selects equivalent residues between the two proteins. The initial set of residue pairs was selected using allowed differences in torsional angles. In the first pass 33% of residue comparisons were selected, whilst for the second pass this reduces to 0.2%. The correct alignment is obtained after the second pass through the program (see Fig. 11).

i) STRUCTURAL ALIGNMENT USING FULL SETS OF RESIDUES

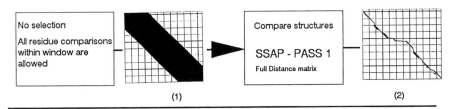

(1) (2)

ii) STRUCTURAL ALIGNMENT USING SUBSETS OF RESIDUES

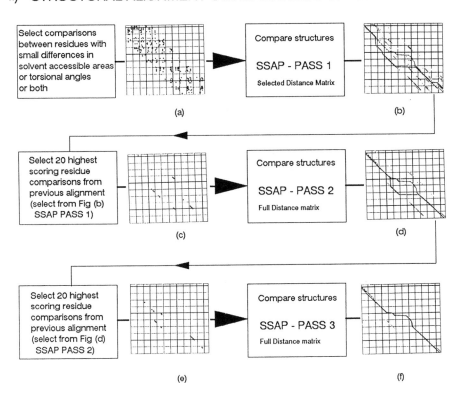

FIGURE 11. Flowchart for the comparison of two structures using the fast alignment method of Taylor and Orengo. Section (i) shows the original method which uses full sets of residue pairs and one pass through the alignment program. Section (ii) describes the three pass method. Comparisons are performed by up to three passes through the structure alignment program using subsets of residue pairs selected by iterative refinement.

Conclusions

There now exists a diverse range of methods available for structure comparison geared towards examining different levels of structural organisation. Some of these accommodate insertions/deletions better and can give good alignments of remote structures, whilst others are more usefully applied to gain a detailed description of similarities and conserved regions in homologous proteins. Several operate at the level of secondary structure matching and allow fast comparisons to be performed, across the databank. Application of these methods should enable different aspects of protein structure to be explored and hopefully yield new understanding of motifs and folding intermediates to improve the prediction of protein structures.

A number of the methods described above, especially those based on secondary structures comparisons are fast enough to perform a detailed analysis of the protein structure databank. The graph theory method of Mitchell *et al.* (1989), the fragment based approach of Rackovsky, (1990) and the latter developments of Taylor and Orengo *et al.* have been specifically developed with the aim of analysing and clustering the databank to identify global similarities and common substructures between proteins. It is hoped that these extensive analyses will reveal new structural motifs.

References

Abagyan, R. A. and Maiorov, V. N. (1988). *J. Biomolec. Struct. and Dynam.*, 5:1267–1279.

Cullis, A. F., Muirhead, H., Perutz, M. F., Rossman, M. G., and North, A. C. T. (1962). *Proc. Roy. Soc.*, A265:161.

Karpen, M. E., Haseth, P. L., and Neet, K. E. (1989). *PROTEINS: Structure, Function and Genetics*, 6:155–167.

Kuntz, I. D. (1975). *J. Am. Chem. Soc.*, 97:4362–4362.

Levine, M., Stuart, D., and Williams, J. (1984). *Acta. Cryst.*, A40:600–610.

Liebman, M. N. (1987). In F., G. J. and Duax, W. D., editors, *Molecular Structure and Biological Activity*, pages 193–212.

Matthews, B. W., Remington, S. J., Grutter, M. G., and Anderson, W. F. (1981). *J. Mol. Biol.*, 147:545–558.

Matthews, B. W. and Rossman, M. G. (1985). *Methods in Enzymology*, 115:397–420.

McLachlan, A. D. (1982). *Acta Cryst.*, A38:871–873.

Mitchell, E. M., Artymiuk, P. J., Rice, D. W., and Willett, P. (1989). *J. Mol. Biol.*, 212:151–166.

Murthy, M. R. N. (1984). *FEBS Letts.*, 168:97–102.

Needleman, S. B. and Wunsch, C. D. (1970). *J. Mol. Biol.*, 48:443–453.

Nishikawa, K. and Ooi, T. (1974). *J. Theor. Biol.*, 48:443–453.

Nishikawa, K., Ooi, T., Ysogai, Y., and Saito, N. (1972). *J. Phys. Soc. Jpn.*, 32(1331–1337).

Nussinov, R. and Wolfson, H. J. (1991). *Proc. Nat. Acad. Sci. USA.* In press.

Orengo, C. A., Brown, N. P., and Taylor, W. R. (1991). In preparation.

Orengo, C. A. and Taylor, W. R. (1990). *J. Theor. Biol.*, 417:517–551.

Padlan, E. A. and Davies, D. R. (1975). *Proc. Natl. Acad. Sci. USA.*, 72:819–823.

Phillips, D. C. (1970). In *Development of Crystallographic Enzymology*, volume 31, pages 11–28. Biochem. Soc. Symp.

Rackovsky, S. (1990). *PROTEINS: Structure, Function and Genetics.*, 7:378–402.

Rackovsky, S. and Scheraga, H. A. (1978). *Macromolecules*, 6:1168–1174.

Rackovsky, S. and Scheraga, H. A. (1980). *Macromolecules*, 13:1440–1453.

Rackovsky, S. and Scheraga, H. A. (1984). *Acc. Chem. Res.*, 17:209–213.

Rao, S. T. and Rossman, M. G. (1973). *J. Mol. Biol.*, 76:241–256.

Remington, S. J. and Matthews, B. W. (1978). *Proc. Nat. Acad. Sci. USA*, 75:2180–2184.

Remington, S. J. and Matthews, B. W. (1980). *J. Mol. Biol.*, 140:77–99.

Remington, S. J., Ten Eyck, L. F., and Matthews, B. W. (1977). *Biophys. Res. Comm.*, 75:265–270.

Richards, F. M. and Kundrot, C. E. (1988). *PROTEINS: Structure, Function and Genetics*, 3:71–84.

Rossman, M. G. and Argos, P. (1975). *J. Biol. Chem.*, 250:7525–7532.

Rossman, M. G. and Argos, P. (1976). *J. Mol. Biol.*, 105:75–96.

Rossman, M. G. and Argos, P. (1977). *J. Mol. Biol.*, 109:99–129.

Schulz, G. E. (1977). *J. Mol. Evol.*, 9:339–342.

Schulz, G. E. (1980). *J. Mol. Biol.*, 138:335–347.

Sippl, M. J. (1982). *J. Mol. Biol.*, 156:359–388.

Sāli, A. and Blundell, T. L. (1990). *J. Mol. Biol.*, 212:403–428.

Sāli, A. Overington, J. P., Johnson, M. S., and Blundell, T. L. (1990). *TIBS*, 15:235–240.

Taylor, W. R. (1988). *J. Mol. Evol.*, 28:161–169.

Taylor, W. R. and Orengo, C. A. (1989a). *J. Mol. Biol.*, 208:1–2.

Taylor, W. R. and Orengo, C. A. (1989b). *Protein Engineering*, 7:505–519.

Ullman, J. R. (1976). *J. Assoc. Comput. Mach.*, 23:31–42.

Zuker, M. and Somorjai, R. L. (1989). *Bulletin of Mathematical Biology*, 51:55–78.

Patterns of Sequence and 3-D Structure Variation in Families of Homologous Proteins: Lessons for Tertiary Templates and Comparative Modelling

Tom Blundell

Imperial Cancer Research Fund Unit of Structural Molecular Biology
Department of Crystallography
Birkbeck College, London University
Malet St.
London WC1E 7HX
U.K.

Introduction

More than 20 years ago X-ray analysis (Adams *et al.*, 1969) showed that insulin was a globular protein which could form dimers and hexamers. Although there were very few sequences of proteins in general available, there were several sequences of insulin (Fig. 1). When these were aligned, it was immediately apparent that much of the pattern of amino acid substitutions could be understood in terms of the three-dimensional structure. For example, the conserved disulphide bridges could be seen to stabilize the association of the A and B-chains and two of them were buried in the core of the protein structure. It was also apparent that many of the amino acid residues in the core were conserved as hydrophobic; for example leucine, isoleucine, valine and phenylalanine. But there was a problem with our structure analysis: residue (A16), an invariant leucine, was fully exposed to the solvent and surrounded by other surface residues that were both polar and variable. As a result of the concern over this structural feature, we were forced to consider remodelling the structure. We discovered that a better interpretation of the electron density could be obtained with a conformation in which the leucine contributed to the conserved hydrophobic core (Blundell *et al.*, 1971).

This experience with the interpretation of the insulin structure convinced me that it is very important to consider all available sequence and chemical information when assessing a new three-dimensional structure. Because the structure as well as the intermolecular interactions provide a set of constraints on the acceptance of amino acid substitutions during evolution, the variation of sequences in an homologous

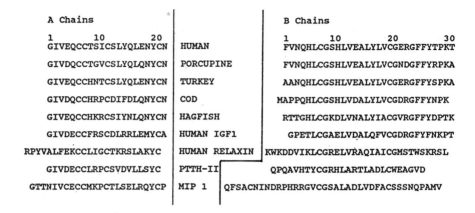

FIGURE 1. The alignment of sequences of insulins (human, porcupine, turkey, cod, hagfish), insulin-like growth factor I (IGF1), relaxin, prothoracicotrophic hormone (PTTH) and mollusk insulin-like peptide (MIP). Numbering as for porcine insulin. Note the conservation of cystines and glycine B10. Residues A2, A6, A7, A11, A16, A19, A20, B11, B15, B18 contribute to the hydrophobic core.

family must have an explanation in the structure and function of the protein. If the explanation is not forthcoming from the structure, then functional interactions might be implied. If this is not likely, the validity of the three-dimensional structure should at least be reassessed. In fact several incorrect X-ray analyses could have been avoided if this procedure had been adopted (see for example, Branden and Jones, 1990).

However, this experience was important in a further way: it clearly demonstrated that the three-dimensional structure implies some knowledge of the sequence variation because amino acid substitutions must be compatible with conservation of the general fold during divergent evolution. It introduced the idea of tertiary templates (to use the modern terminology) as a set of restraints applied from the three-dimensional structure level on to the sequences.

In the following years we used this approach to identify other more distant members of the insulin family. In each case we sought an invariant pattern of cystines and glycines (which allow sharp turns in the polypeptide chain) and a conservatively varied pattern of residues that would comprise the hydrophobic core. For each candidate sequence we modelled the three-dimensional structure (using plastic or wire models at first) to test the hypothesis that it formed a similar three-dimensional structure. Between 1971 and 1975 I built almost all of the unusual hystricomorph or non-mammalian sequences as three-dimensional models (see Blundell and Wood, 1975 for review). In 1976 we discovered the sequence of another hormone, relaxin, that had the same pattern of conserved residues and showed that although the percentage sequence identity was low (about 20%; see Fig. 1), the protein could adopt an insulin-like conformation (Bedarkar et al., 1977), Currently, we have around 65 insulin sequences including insulins proper, the insulin-like growth factors (Blundell et al., 1978; Blundell and Humbel, 1980), insulin-like peptides from the silk worm (Jhoti et al., 1987) and mollusks (Murray-Rust et al., 1991). In all cases the cystines are invariant and the B8 Gly is conserved so that it can assume a positive-ϕ main-chain torsion angle, and, with one exception in the mollusk insulin-like peptides, the core is conserved as hydrophobic.

In this article I consider what can be learnt from high resolution X-ray crystallographic studies of proteins from comparisons of the sequences and three-dimensional structures of families of homologous proteins. My discussion will be largely historical from the point of view of the structure analyses in which I have participated; the proteins include families of lens crystallins and aspartic proteinases. The emphasis of our work has been to learn rules from careful comparisons and then to quantify and automate the procedures so that they can be applied to protein structure more generally. Thus we have developed algorithms for comparisons of both sequences and three-dimensional structures (Sutcliffe et al., 1987; Säli and Blundell, 1990). We have developed rules in terms of substitution tables (Overington et al., 1990) or more generally of multidimensional probability tables (Säli, 1990). We have used these rules in the automatic development of sequence templates that characterize a protein fold for an homologous family (Overington et al., 1990; M.Johnson unpublished results) and three-dimensional models that define a particular three-dimensional structure (Sutcliffe et al., 1987; Blundell et al., 1988; Säli et al., 1990).

X-Ray Analyses of Families of Proteins

Although much information can be gained from the study of a large number of sequences in conjunction with a single three-dimensional structure of a protein such as insulin, there are many uncertainties in this procedure. These relate to the difficulties in making sequence alignments that are meaningful in structural terms, especially if the sequence identities are less than 40%. Furthermore, regions that are aligned optimally by sequence may still have quite different local conformations and tertiary interactions. For these reasons it is desirable to consider families of proteins for which several three-dimensional structures are available.

In order to illustrate this point I shall describe two families of proteins, the three-dimensional structures of which have been determined at Birkbeck and which influenced our own approach to the development of automated procedures for protein comparisons. In each of these families we have become familiar with the details of the structure and function through biochemical and X-ray analyses. The protein families are the β/γ crystallins and the aspartic proteinases.

The β/γ crystallins comprise a family of homologous proteins, the protomers of which are composed structurally of four Greek key motifs arranged as two globular domains. X-ray analyses at high resolution have defined the three-dimensional structures of four closely related monomeric γ crystallins and one more distantly related oligomeric β crystallin (Blundell *et al.*, 1981; Sergeev *et al.*, 1988; Bax *et al.*, 1990). There are twenty structures for the Greek key motifs defined by X-ray analysis and more than 120 sequences of the motifs from homologous proteins available in sequence data bases. The individual motifs are not closely related. Only one residue, a glycine (Gly 13) is identical in all structures (Fig. 2). At an early stage we realised that this glycine must be conserved as a result of the sharp fold in the central β-hairpin of the Greek key motif that was achieved by having the glycine with a positive ϕ torsion angle (Fig. 3). One further residue, a serine (Ser v34), is conserved in most sequences but varied occasionally to alanine. This is also important to the three-dimensional structure as it is buried and forms hydrogen bonds through its sidechain oxygen to nearby mainchain NH and CO functions. The sequences of several motifs from β and γ crystallins are shown in Fig. 2.

The aspartic proteinases include the pepsins, which are bilobal enzymes with a deep and extended active site cleft. The two catalytically active aspartates lie in conserved sequences (Asp-Thr-Gly) at the centre of the cleft. These catalytic aspartates (Asp 32 and Asp 215 in pepsin) occupy equivalent positions on the two lobes, which have little other sequence identity but have topologically similar structures (Tang *et al.*, 1978). Our detailed analyses at Birkbeck are of the structures of the mammalian enzymes, chymosin (Newman, 1990; Strop *et al.*, 1991) and pepsin (Cooper *et al.*, 1990), and two fungal enzymes, endothiapepsin (Blundell *et al.*, 1990) and mucorpepsin (Newman, 1990). These enzymes have between 25% and 60% identities when considered pairwise. Other three-dimensional structures of aspartic proteinases available from the Brookhaven Databank include the fungal enzymes penicillopepsin (James and Sielecki *et al.*, 1983) and rhizopuspepsin (Suguna *et al.*, 1987) and independent structures of

```
              10            20            30            40          50
cfe1      G k I t F ȳ e ā̆ r g ƒ q̲ g r h y ē C s̃ - s d h s n L q p y - - - - - F s r̲ C n̰ S̲ I r̃ V d s
ratea1    G k I t F y e ā̆ r g ƒ q̲ g r h y e C s - s d h s ñ L q̄ p y - - - - - F s r̲ C n S̲ I r V d s
cfrb1     g k I t F y ē ā̆ r g ƒ q̲ g h̲ c y e C s - s d c p n L q p y - - - - - F s̃ r̲ C n S̲ I r V d s
cfbb21    l n p k I i I f e q e n ƒ q g h̲ s h ē l n̰ - g p c p ñ L k̃ e T - - - - g V e k A g S̲ V l V q a
cfe3      H ĩ L ĩ I y ē r e d y̲ r g q̃ m v ē I t̲ - ē d C s s̃ L q ā̆ ĩ - - - f h ƒ s d I h̲ S̲ F h V m e
ratea3    h ĩ L ĩ I y ē r e d y̲ r g q̃ m v ē I t̲ - ē d C s s̃ L q ā̆ ĩ - - - f h ƒ s d I h̲ S̲ F h V m e
cfrb3     F ĩ M ĩ I y e ā̆ d ƒ ĩ̲ g q m s ē I t - ā̆ d C p s̃ L q ā̆ ĩ̲ - - - f h̆ l t e V h̲ S̲ L n V l e
cfbb23    h̃ k̃ I t L y e ñ p n ƒ t̲ g k̃ k m ē V i ā̆ d d V p s F h a h̲ - - - g Y̲ q e k̃ V S̲ S̲ V r V q s
cfe2      g c WML Ŷ e q p n ƒ t̲ g c q YFL r r g d y̲ p ā̆ y q q̃ w m - - G f s d s̲ V ĩ S̲ C ĩ l I p h t s
ratea2    g c WML Ŷ e q p n ƒ t̲ g c q YFL r r g d y̲ p d y q q W m - - G f s d s̲ V ĩ̲ S̲ C ĩ l I p h t s
cfrb2     g c WML Ŷ e r p n y q̲ g h̃ q YFL r ĩ g ā̆ y̲ p ā̆ y q̃ q W m - - G f n d s̲ I ĩ S̲ C ĩ l I p q h̲ t
cfbb22    G p Wv G ȳ e q a n c k g ē q F v F e k g e y̲ p r w d s̲ W T̲ s s̃ r̲ ĩ t d s L s̲ S̲ L ĩ p i k̲ v ā̆ s
cfe4      g y WV L Ŷ e m p n y r g r Q y L L r p g d y̲ r r y l d W g̲ - - A a ñ a ĩ V g S̲ L ĩ r A v d f y
ratea4    G y WV L Ŷ e m p n y r g r Q̃ Y L L r p g d y̲ r r y l d W g - - A a ñ a ĩ V g S̲ L ĩ r A v d f y
cfrb4     g S WV L Ŷ e m p s y r̲ g ĩ Q̃ Y L L r p g ē y̲ ĩ r y l d W g - - A m n A k V g S̲ L ĩ r V m d f y
cfbb24    g t Wv G Ŷ q y p g y r̲ g l q Y l L e k g d y̲ k d s g d F g - - A p q̃ p q̃ V q̲ S̲ V ĩ r i r̲ d m q

          β β β β β β    + + β + β β β β β                            β β β β
```

FIGURE 2. The alignment of sequences of the four Greek key motifs of three γ and one β crystallin obtained by comparing their 3-D structures using COMPARER. For each sequence there are four motifs. The numbering is that of the first motif of γB. The amino acid code is the standard one-letter code formatted using the following convention (Overington *et al.*, 1990): *Italic* for positive φ; UPPER CASE for solvent inaccessible; lower case for solvent accessible; **bold** for hydrogen bonds to mainchain amide nitrogen; underline for hydrogen bonds to mainchain carbonyl oxygen; tilde ~ for sidechain-sidechain hydrogen bonds. The secondary structure is given below where it is present in 80% or more of the proteins. a: α-helix; B: β-strand; +: positive φ torsion angle.

Key: cfe = caffe; ratea = Rat E (A chain); cfrb = calf γB; cfbb2 = calf βB2.

Motif 1	φ ψ Acc	Motif 3	φ ψ Acc	Motif 2	φ ψ Acc	Motif 4	φ ψ Acc	Conf
1	—	84	−96	40A	69	123	73	
Gly	−166	Phe	136	Gly	−179	Gly	168	β
1	37	88	4	40	12	129	6	
2	−143	85	−143	41A	−104	124	−104	
Lys	134	Arg	125	Cys	132	Ser	149	β
2	90	89	80	41	12	130	10	
3	−149	86	−144	42A	−130	125	−137	
Ile	140	Met	135	Trp	158	Trp	162	β
3	0	90	1	42	0	131	0	
4	−131	87	−109	43A	−133	125A	−116	
Thr	132	Arg	134	Phe?	112	Val	123	β
4	23	91	.75	43	0	132	0	
5	−107	88	−116	44A	−101	125B	−104	
Phe	138	Ile	145	Ile?	148	Leu?	124	β
5	0	92	2	44	0	133	1	
6	−127	89	−121	45A	−127	125C	−103	
Tyr	141	Tyr	151	Tyr	150	Tyr	139	β
6	33	93	24	45	3	134	7	
7	−72	90	−62	47A	−79	126	−56	
Glu	−42	Glu	−31	Glu	−29	Glu	−49	α
7	17	94	94	46	42	135	36	
8	−108	91	−122	48A	−112	127	−117	
Asp	169	Arg	177	Met?	149	Met	170	β
8	48	95	99	47	99	136	60	
9	−74	92	−92	49	−61	128	−62	
Arg	−176	Asp	152	Pro	158	Pro	58	Col
9	149	96	62	48	50	137	25	
10	52	93	74	50A	56	129	64	
Gly	16	Asp	31	Asn	52	Ser	50	αL
10	37	97	75	49	81	138	62	
11	70	94	55	51	52	130	46	
Phe	42	Phe	52	Tyr	59	Tyr	56	αL
11	54	98	37	50	45	139	56	
12	−113	95	−122	51A	−133	131	−134	
Gln	155	Arg	168	Gln	164	Arg	154	β
12	117	99	96	51	124	140	122	
13	99	96	75	52	74	132	87	
Gly	−162	Gly	−167	Gly	−162	Gly	−156	Gly
13	50	100	30	52	23	141	27	
14	−67	97	−73	53	−56	133	−73	
His	142	Gln	147	His	139	Arg	142	Col
14	94	101	124	53	41	142	80	
15	−134	98	−119	54A	−118	134	−114	
Cys	161	Met	149	Gln	136	Gln	145	β
15	65	102	74	54	22	143	12	

(bracket *a* spans rows 1–6; bracket *b* spans rows 14–15)

FIGURE 3. The sequences, accessible surface areas and conformational parameters for the four motifs of γB crystallin (cf. Fig. 2) (by permission of Dr. G. Wistow).

pepsin (Andreeva *et al.*, 1984; Sielecki *et al.*, 1990) and chymosin (Gilliland *et al.*, 1990). The aspartic proteinases also include the retroviral proteinases. Structures have been determined for proteinases from Rous Sarcoma Virus (RSV; Miller *et al.*, 1989) and Human Immunodeficiency Virus (HIV; Wlodawer *et al.*, 1989; Lapatto *et al.*, 1989). In these dimeric enzymes each subunit corresponds to a lobe of the pepsins and contributes one catalytic aspartate within a conserved sequence of Asp-Thr/Ser-Gly. Only three residues are identical in all lobes/subunits of aspartic proteinases. These include the aspartate (Asp32 and equivalents) and glycine of the sequences at the active sites and a further glycine (Gly 123 and equivalents) in a strand that is close-by. These glycines are conserved not because they have an unusual mainchain torsion angle but because sidechains would disrupt the three-dimensional structure particularly of the catalytic residues. The second residue in the catalytic triad is usually conserved as threonine but is occasionally varied to serine; like the conserved serine in crystallins, this residue is completely buried in the core and hydrogen bonds to nearby mainchain NH and CO functions. Sections of the sequences of these aspartic proteinases in the conserved regions are shown in Fig. 4.

Rules Expressed as Environment-Specific Substitution Tables

Our own experimental work and its analysis gave further weight to the idea that particular substitution patterns may result from structural constraints within the molecule except where the amino acids interact with substrate or other molecules important to the function. For example, solvent-inaccessible residues, whose sidechains give a close-packed core, have a lower rate of acceptance and a more limited set of substitutions than those on the surface [see also for example Chothia and Lesk, 1986; Hubbard and Blundell, 1987]. Inter-residue hydrogen bonds, especially with peptide NH or CO functions, can also act as a constraint on the substitution of amino acids (Bajaj and Blundell, 1984; Blundell, 1986). Secondary structure provides further strong constraints on sequence variability; in particular residues with a positive ϕ torsion angles have characteristic substitution patterns.

We have attempted to characterize and quantify this information concerning the evolution of proteins (Overington *et al.*, 1990). Our analysis depends on a systematic approach to the comparison of sequences and three-dimensional structures using COMPARER (Săli and Blundell, 1989; Zhu *et al.*, 1990). COMPARER leads to an

```
HIV     q l K ĝ A L L D̃ T̃ G A  d̃ d T̃ V L e ẽ - - - - - - - - - - - - - - - M s L p - -
2RSV    V y I t A L L D̃ S G A D̃ I T̃ I I Š e e d̃ WP - - - - - - - - - - - t d WP - -
                30          40          50
              30          40          50          60

4APE-N  g̃ ĩ L n L D F D̃ T̃ G Š  S̃ D̃ L W̃ V F Š s ẽ T̃ ι a - - s e v d g Q̃ t i Ỹ T̃ P s k Š
2APP-N  t ĩ L n L N̲ F D̃ T̃ G s̃  A D̃ L W V F S̲ ĩ ẽ L p a - - s q q s g H̃ s V Ỹ ñ P š a ĩ
2APR-N  k ǩ F ñ L D F D̃ T̃ G S̲ S̃ D̃ L W̃ I A S t 1 C ĩ ñ - - C - g s g Q̃ t k Ỹ d P n q Š
PEP-N   q d F t V I F D̃ T̃ G S  S̃ Ñ L W̃ V P S̲ v y C s s 1 A C - - s d H̃ ñ q F ñ P d̃ d Š
CHY-N   q ẽ F T V L F D̃ T̃ G s̃  S̃ D̃ F W V P S̲ I y C k Š n A C - - k n H̃ q r F D̃ P f̃ k Š
4APE-C  t s I d̃ G I A D̃ T̃ G ĩ  ι L L y L p - - - - - - - - - - - a t V V s a Y̲W̲ a q V
2APP-C  d G f s G I A D̃ T̃ G ĩ  ι L L 1 L d̃ - - - - - - - - - - d s V V s q̃ Y̲Y̲ s q V
2APR-C  s s F d̃ G I L D̃ T̃ G ĩ  ι L L i L P - - - - - - - - - - ñ n i A a s V A r a Y̲
PEP-C   g g C q̃ A I V D̃ T̃ G ĩ  s 1 L T G P - - - - - - - - - - ĩ s a l a n I Q̃ s d I
CHY-C   g G c q A I L D̃ T̃ G t  s k L V G p - - - - - - - - - - s s d I 1 n I Q q a I
        β β β β β β β             β β β                              α
```

```
HIV     - - p v N I I G - - - - - - - - - - - - - R̃ ñ L L T̲ q I
2RSV    - - r g Š I L G - - - - - - - - - - - - - R̃ d̲ C L q g L
                        110
                      120         130         140

4APE-N  s ĩ I D G L L G L A f s̃ t 1 Ñ t̲ V s p t q q k T F F d̃ ñ A
2APP-N  t ñ Ñ D̃ G L L G L A F s̃ s i Ñ t̃ V q p q̃ s q ĩ T F F d̃ ĩ V
2APR-N  - P Ñ D G L L G L G F d̃ t i T̃ ĩ V r - - g V k T̃ P M d̃ Ñ L
PEP-N   - p F D̃ G I L G L A Ỹ p s i Š a s - - - g A t P V F D̃ Ñ L
CHY-N   - s F D G I L G M A Ỹ p s l A s̃ e - - - y Š i P V F D̃ Ñ M
4APE-C  - - g i Ñ I F G - - - - - - - - - - - - - D̲ V A L Ǩ A A
2APP-C  - - g f S I F G - - - - - - - - - - - - - D̲ I F L Ǩ S Q̃
2APR-C  - w g F A I I G - - - - - - - - - - - - - D̲ T̲ F L Ǩ Ñ N̲
PEP-C   s g ẽ L W I L G - - - - - - - - - - - - - D̲ V F I R̃ q Y̲
CHY-C   - - q k W i L G - - - - - - - - - - - - - D V F I R̃ Ẽ Y
        β β β                                     α α
```

FIGURE 4. A section of the alignment of sequences of aspartic proteinases achieved by comparing the three-dimensional structures using COMPARER [Săli and Blundell, 1990]. APE: endothiapepsin; APP: penicillopepsin; APR: rhizopuspepsin; PEP: hexagonal porcine pepsin; CHY: calf chymosin; RSV: Rous sarcoma virus proteinase; HIV: human immunodeficiency virus proteinase. The last letter refers to the amino (N) or carboxy (C) terminal domains of the pepsins. One letter code as in Fig. 2.

alignment of the sequences based upon the equivalence of the structures locally. COMPARER has been used to compare and align families of proteins for which there are several high resolution X-ray analyses and coordinates in the Brookhaven Protein Databank (Bernstein *et al.*, 1977). The alignments of the crystallins and aspartic proteinases shown in Figs. 2 and 4 have been produced using this approach. The structural features of each amino acid are also represented in these figures.

In the calculation of substitution tables (Overington *et al.*, 1990), we first considered the structural features that appeared to be important in the crystallin and aspartic proteinase families, which have been studied experimentally at Birkbeck. These features were residue type (20 values), accessibility (2 values), side chain hydrogen bonding (8 values) and main chain conformation (4 values). This gave a maximum of $20 \times 2 \times 8 \times 4$ classes of amino acids. In fact several amino acids are unable to form hydrogen bonds through their sidechains and most polar residues are unable to act both as donors and acceptors except at extreme pH values. Furthermore inaccessible ion pairs rarely occur except at domain or subunit interfaces, but these were largely omitted from the study. As a result of these factors the effective number of classes was about 300.

All pairwise comparisons of structures in each alignment produced by COMPARER were considered in the analysis, and all substitutions implied by pairwise comparisons were stored in tables as a function of the features identified in the three-dimensional structures. In order to avoid very sparse tables, we considered the structural features of only one of the two proteins compared. It is convenient to display the data as 20 by 20 matrices where one dimension refers to the amino acid type restricted to a stuctural environment and the other is simply residue type. Fig. 5 shows some columns from the substitution tables for solvent inaccessible residues that are hydrogen-bonded to mainchain NH functions.

Fig. 5 shows that the largest value for conservation is seen for aspartic acid (Fig. 5a), which exceeds others attributable to hydrogen bond interactions. On the relatively infrequent occasions when substitutions are accepted at such positions, an asparagine or serine, which have similar hydrogen bonding capacity, are most likely to occur. This contrasts strongly with the substitution patterns of asparagine (Fig. 5b). Inaccessible asparagines with sidechain to mainchain NH hydrogen bonds are substituted more often with aspartate or serine than with asparagine; leucines, alanines and many other residues are accepted. Surprisingly glutamine differs greatly from asparagine but resembles aspartate in its relatively high conservation. Its substitution profile indicates that glutamate and histidine are preferred substituents. Similar strong preferences for conservation are shown by solvent inaccessible serine and threonine. These show that

198

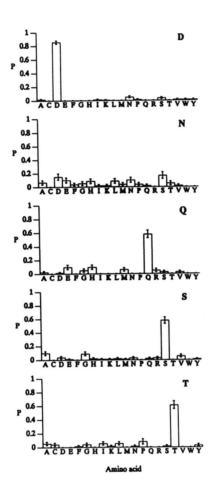

FIGURE 5. Patterns of substitution for amino acids that are solvent inaccessible and hydro-gen-bonded to mainchain NH for (a) Asp, (b) Asn, (c) Gln, (d) Ser, (e) Thr. Probabilities (P) of a given residue being replaced by any of the 20 amino acids are given with standard errors.

the conservation of the buried serines and threonines in the crystallins and aspartic proteinases are characteristic of a more general trend in the evolution of protein structure. Some examples of conserved residues taken from structures defined by X-ray analysis are given in Fig. 6.

The existence of a positive ϕ also puts strong constraints on the variation of an amino acid. If the values ϕ and ψ are in the region of a left handed α-helix, such as Gly 10 of the crystallins, then the glycine is relatively well conserved but can be substituted mainly by aspartic acid, asparagine or sometimes serine (see Figs. 2 and 3). For other positive values of the torsion angle ϕ, there are strong constraints for the conservation of glycine; an example is the uniquely conserved glycine (Gly 13; Figs. 2 and 3) that facilitates the folding of a β-hairpin onto the β sheet formed from Greek key motifs in the crystallins.

Rules for Generating Tertiary Templates

The structural data, comparisons and substitution tables provide a quantitative analysis of amino acid diversity in homologous proteins. We have already shown that they can be used predictively to estimate the probable amino acid variation at each position in a protein of known three dimensional structure (Overington et al., 1990; M. S. Johnson, unpublished results). This provides a general approach to constructing templates on the basis of the tertiary structure. For each topologically equivalent position in each known structure, we use the tables to predict the substitution of amino acid residues. Likely places for insertions and deletions can also be predicted. The method is complementary to the approach of Ponder and Richards [1988].

Templates constructed on the basis of one or more 3-D structures are complementary to those constructed from the alignment of many sequences (Taylor, 1986). Both kinds of template can define sequence fingerprints that are essential to structure or function. They can be used in the form of consensus sequences or substitution tables to search out distantly related proteins in the sequence database.

The templates of all known three-dimensional structures or families of structures including loops, motifs, domains and complete globular proteins should be precalculated so that a new sequence can be compared with them rather than with individual proteins. This will result in a better alignment of whole proteins or their parts and thereby in a better extrapolation of spatial features from known structures onto the sequence of the unknown in knowledge-based modelling.

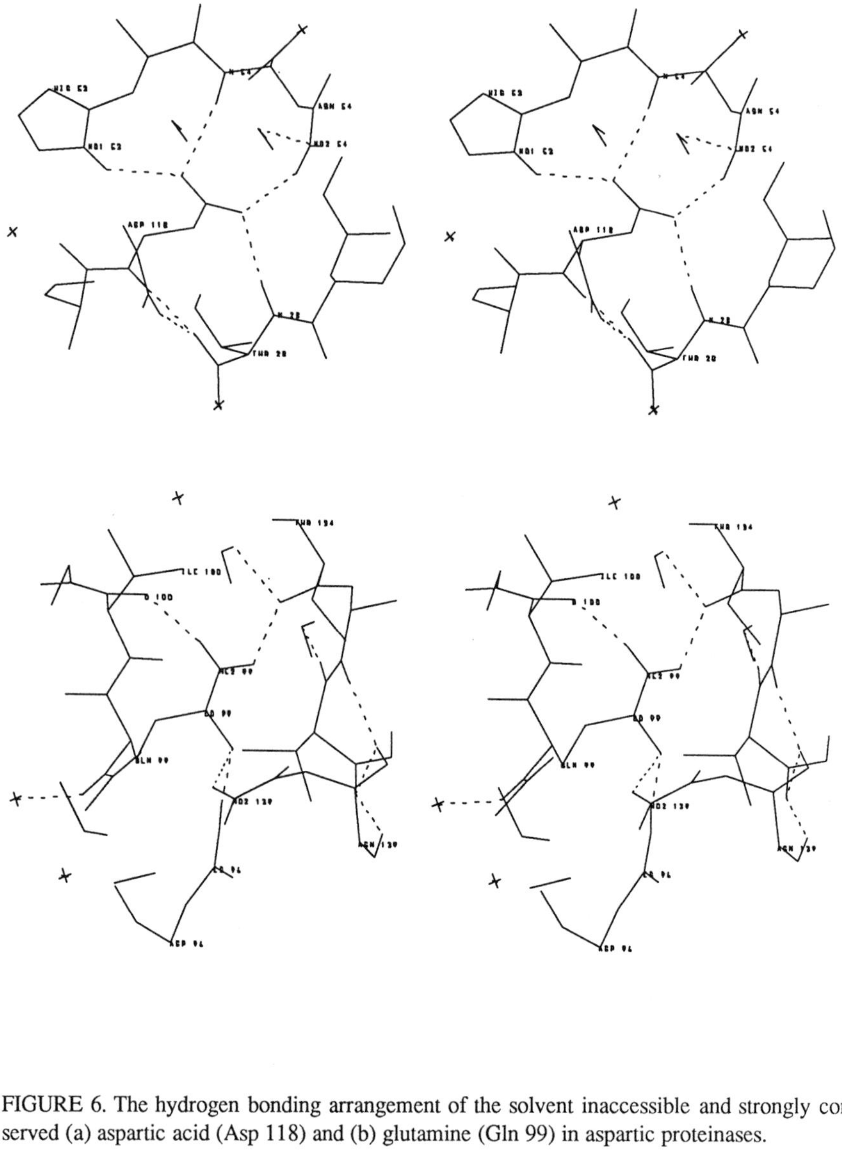

FIGURE 6. The hydrogen bonding arrangement of the solvent inaccessible and strongly conserved (a) aspartic acid (Asp 118) and (b) glutamine (Gln 99) in aspartic proteinases.

Rules for Modelling Three-Dimensional Structure

We have shown that analyses of aligned structures of homologous proteins can give rise to simple rules. These rules can then be used in the construction of a three-dimensional model from a sequence. I shall not describe these in detail here; the reader may refer to the recent review of Såli *et al.* (1990) and the references therein.

Most approaches depend on the assembly of fragments of three-dimensional structures (Greer, 1981; Jones and Thirup, 1986; Blundell *et al.*, 1988). For example in the computer program COMPOSER (Sutcliffe *et al.*, 1987; Blundell *et al.*, 1988), we select three sets of fragments. The first set is derived from the framework defined by multiple superposition of the chosen homologous structures. A second set of protein fragments for regions outside the framework is selected from the database of loop sub-structures using a distance filter in a similar way to Jones and Thirup (1986). The third set of rigid fragments, the sidechains, is selected using rules derived from the analysis of homologous structures. The 1200 rules derived include one for each of the 20 by 20 amino acid replacements in each of the three secondary structure types (α-helix, β-strand or irregular). The templates of selected fragments are clustered and ranked using the methods described above,and the top ranking fragments are annealed together. The model is checked for serious overlaps between fragments; where this occurs the next ranking fragment is used. The final model is energy minimised to remove minor inconsistences.

This modelling procedure is very successful where the known structures cluster around that to be predicted and where the percentage sequence identity to the unknown is high (greater than 40% identity). In all cases the accuracy of the prediction decreases very quickly as the sequence identity between the known and unknown decreases. For these cases a new modelling technique is required that is not restricted by the idea of assembling rigid fragments of protein structure. The best procedure is based on distances between atoms. This is similar to methods for structure analysis using 2D-NMR data.

In our procedure (Såli, 1990; Såli *et al.*, 1990), we first use the alignment of the sequence to be modelled with sequences of known related structures to derive a list of distance constraints. Tables derived from the comparisons of homologous proteins are used to predict mainchain and sidechain dihedral angles, $C\alpha - C\alpha$ distances and hydrogen-bonding distances from the known structures aligned with that being modelled. The predicted distances are expressed as Gaussian probability functions. For non-bonded atoms this often involves taking a mean distance from an homologous protein and a standard deviation proportional to the similarity between the proteins and the magnitude of the distance. For sidechain dihedral angles the probability functions are usually trimodal with the relative magnitudes depending on the particular residue type and the values of equivalent dihedral angles in related known structures. In general, every structural feature can be constrained by several sources. The three-dimensional model is constructed to satisfy these constraints. We have constructed models of several proteins using this procedure (Sāli, 1990); the first results look very encouraging.

Acknowledgements: I am very grateful to my colleagues at Birkbeck who have been involved in both the experimental and computational aspects of this work. They include Jon Cooper, Dan Donnelly, Huub Driessen, Yvonne Edwards, Frank Eisenmenger, Carlos Frazao, Mark Johnson, Alasdair McLeod, Matthew Newman, Karsten Niefind, John Overington, Andrej Sāli, Lynn Sibanda, Christine Slingsby, Pam Thomas, Janet Thornton, Ian Tickle, Chris Topham, Nalini Veerapanini and Shangyang Zhu for many stimulating discussions.

References

Adams, M., Blundell, T. L., Dodson, E., Dodson, G. G., Vijayan, M., Baker, E. N., and Hodgkin, D. C. (1969). *Nature*, 224:491–495.

Andreeva, N. S., Zdanov, A. S., Gustchina, A. E., and Federov, A. A. (1984). *J. Mol. Biol.*, 259:11353–11365.

Bajaj, M. and Blundell, T. L. (1984). *Ann. Rev. Biophys. Bioeng.*, 13:453–492.

Bax, B., Lapatto, R., Nalini, V., Driessen, H., Lindley, P. F., Mahadevan, D., Blundell, T. L., and Slingsby, C. (1990). *Nature*, 347:776–780.

Bedarker, S., Turnell, W. G., Schwabe, C., and Blundell, T. L. (1977). *Nature*, 270:449–451.

Bernstein, F. C., Koetzle, T. F., Williams, G. J. B., Mayer, E. F., Brice, M. D., Rogers, J. R., Kennard, O., Shimanouchi, T., and Tasumi, N. (1977). *J. Mol. Biol.*, 112:535–542.

Blundell, T. L. (1986). *Chemica Scripta*, 26B:213–219.

Blundell, T. L., Bedarkar, S., Rindernecht, E., and Humbel, R. E. (1978). *Proc. Nat. Acad. Sci. USA*, 75:108–184.

Blundell, T. L., Carney, D., Gardner, S., Hayes, F., Howlin, B., Hubbard, T., Overington, J., Singh, D. A., Sibanda, B. L., and Sutcliffe, M. (1988). *Eur. J. Biochem.*, 172.

Blundell, T. L., Cutfield, J. F., Cutfield, S. M., Dodson, E., Dodson, G. G., Hodgkin, D. C., Mercola, D., and Vijayan, M. (1971). *Nature*, 231:506–511.

Blundell, T. L. and Humbel, R. E. (1980). *Nature*, 287:781–787.

Blundell, T. L., Jenkins, J. A., Sewell, B. T., Pearl, L. H., Cooper, J. B., Wood, S. P., and Veerapandian, B. (1990). *J. Mol. Biol.*, 211:919–941.

Blundell, T. L., Lindley, P., Miller, L., Moss, D. S., Slingsby, C., Tickle, I. J., Turnell, W. G., and Wistow, G. (1981). *Nature*, 289:771–777.

Blundell, T. L. and Wood, S. P. (1975). *Nature*, 257:197–203.

Branden, C.-I. and Jones, T. A. (1990). *Nature*, 343:687–689.

Chothia, C. and Lesk, A. M. (1986). *EMBO J.*, 5:823–826.

Cooper, J. B., Kahn, G., Taylor, G., Tickle, I. J., and Blundell, T. L. (1990). *J. Mol. Biol.*, 214:199–222.

Gilliland, G. L., Winbourne, Y. L., Nachman, J., and Wlodawer, A. (1990). *Proteins*, 8:82–101.

Greer, J. (1981). *J. Mol. Biol.*, 153:1027–1042.

Hubbard, T. and Blundell, T. L. (1987). *Protein Engineering*, 1:159–171.

James, M. N. G. and Sielecki, A. R. (1983). *J. Mol. Biol.*, 163:299.

Jhoti, H., McLeod, A. N., Ishizaki, H., Nagasawa, H., and Suzuki, A. (1987). *FEBS Let.*, 219:419–425.

Jones, T. A. and Thirup, T. (1986). *EMBO. J.*, 5:819–822.

Lapatto, R., Blundell, T. L., Hemmings, A., Overington, J., Wilderspin, A., Wood, S. P., Merson, J. R., Whittle, P. J., Danely, D. E., Geoghegan, K. F., Hawrylik, S. J., Lee, S., Scheld, K. G., and Hobart, P. M. (1989). *Nature*, 342:299.

Miller, M., Jaskolski, M., Rao, J. K. M., Leis, J., and Wlodawer, A. (1989). *Nature*, 337:576–579.

Murray-Rust, J., McLeod, A., Blundell, T. L., Smit, G., and Joosse, J. (1991). Manuscript in preparation.

Newman, M. (1990). PhD thesis, University of London.

Overington, J., Sâli, A., Johnson, M., and Blundell, T. L. (1990). *Proc. Roy. Soc. Lond. B*, 241:132–145.

Ponder, J. W. and Richards, F. M. (1987). *Proteins*, pages 775–791.

Sergeev, Y. V., Chirgadze, Y. N., Mylvaganam, S. E., Driessen, H., Slingsby, C., and Blundell, T. L. (1988). *Proteins*, 4:137–147.

Sielecki, A., Federov, A. A., Boodhoo, A., Andreeva, N., and James, M. N. G. (1990). *J. Mol. Biol.*, 214:143–170.

Strop, P., Newman, M., and Blundell, T. L. (1991). *Biochemistry*. In press.

Suguna, K., Padlan, E. A., Smith, C. W., Carlson, W. D., and Davies, D. (1987). *Proc. Natl. Acad. Sci. USA*, 84:7009–7013.

Sutcliffe, M. J., Heneef, I., Carney, D., and Blundell, T. L. (1987). *Protein Engineering*, 1:377–384.

Săli, A. and Blundell, T. L. (1990). *J. Mol. Biol.*, 212:403–428.

Săli, A., Overington, J. P., Johnson, M. S., and Blundell, T. L. (1990). *Trends in Biochemical Sciences*, 15(6):235–239.

Tang, J., James, M., Sielecki, A., Jenkins, J. A., and Blundell, T. L. (1978). *Nature*, 271:618–621.

Taylor, W. R. (1986). *J. Mol. Biol.*, 188:233–254.

White, H. E., Driessen, H. P. C., Slingsby, C., Moss, D. S., and Lindley, P. F. (1989). *J. Mol. Biol.*, 207:217–235.

Wlodawer, A., Miller, M., Jaskolski, M., Sathyanarayana, B., Weber, I., Selk, L., Clawson, L., Schneider, J., and Kent, S. (1989). *Science*, 245:616–621.

Zhu, Z., Săli, A., and Blundell, T. L. (1990). Unpublished results.

Discussion

Q: It would be nice to know the reason for some of the unexpected substitution patterns, for example, those of asparagine and aspartic acid.

A: They are certainly fascinating. For example why are buried asparagines less conserved than buried aspartates, which are usually charged? On the other hand neutral glatamine tends to be conserved when it is solvent inaccessible. I have yet to find a convincing explanation for these differences. But our data certainly pose some interesting theoretical questions.

Q: When you compile your replacement statistics to make your mutability matrices, do you correct for functionally conserved residues?

A: No, we do not. We hope that they are sufficiently few not to cause too much noise in the data. It is very interesting to look at a protein to see which residues have different substitution patterns than predicted. These are often catalytic or other functional residues.

Q: Would it not be better to include only the well aligned regions in the calculation of the substitution tables?

A: We did discuss this possibility but decided that the approach should approximate the situation of the applications. In real alignments and real modelling we would be dealing with regions of poor homology.

Modelling From Remote Sequence Similarity — Enveloped Virus Capsid Structure Modelled on the Non-Enveloped Capsid

Stephen D. Fuller and Terje Dokland

Biological Structures and Biocomputing Programme
European Molecular Biology Laboratory
Heidelberg
F.R.G.

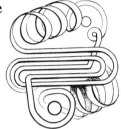

Introduction

A fundamental classification of viruses is a division into enveloped viruses which contain a lipid bilayer and non-enveloped viruses which have only a protein shell. This classification is a particularly frustrating one for the structural biologist since almost all of the information which we have about virus structure concerns non-enveloped viruses. For non- enveloped viruses such as tomato bushy stunt, southern bean mosaic and the picornaviruses we have very detailed information in the form of high resolution structures of the whole virion obtained through X-ray crystallography (Harrison *et al.*, 1978; Hogle *et al.*, 1985; Rossmann *et al.*, 1985; Luo *et al.*, 1987; Acharya *et al.*, 1989). For the structural proteins of enveloped viruses, we have high resolution structures of the soluble portions of two envelope proteins of one virus, influenza (Wilson *et al.*, 1981; Varghese *et al.*, 1983). No direct high resolution information exists for the nucleocapsid underlying the membrane in which the genome of the virus is packaged and whose interactions with the envelope proteins mediate the assembly of the enveloped virion.

Here we describe the use of a combination of electron microscopy, site-directed mutagenesis and computer sequence alignment to draw some conclusions about the structures of enveloped virus nucleocapsids from what we know about non-enveloped viruses. Our two major collaborators have been Dr. Patrick Argos at EMBL, with whom the first sequence alignments for Semliki Forest virus and hepatitis B virus were performed and Dr. Michael Nassel at the Zentrum für Molekularbiologie in Heidelberg, with whom the mutagenesis work on the hepatitis B capsid was performed.

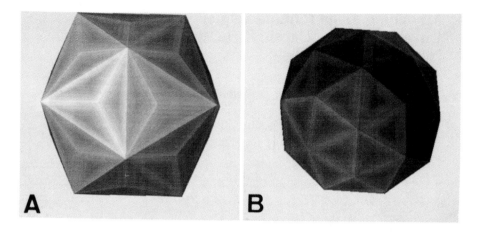

FIGURE 1. (A)Icosahedron with T=1 pattern (B) Icosahedron with T=3 pattern. The 15 two-fold, 10 three-fold and 6 five-fold axes are apparent in each although the T=1 example has 60 identical structural units and the T=3 has 180 similar structural units.

Virus Structure

Many of the enveloped viruses as well as the non-enveloped viruses display icosohedral symmetry. Fig. 1A shows a diagram of an icosahedron. It has 20 triangular faces, 12 corners and 30 edges. This means that the icosahedron consists of 60 identical fundamental units, in this case a stippled chevron. Since viral capsid proteins tend to be of approximately the same size (giving a shell thickness of ~ 40Å), the only way to produce a larger virus is by increasing the number of subunits that make up each fundamental unit; each subunit cannot be exactly identical, since each will have a different environment. The small satellite tobacco necrosis virus is able to package its incomplete genome in an ~ 200Å shell built from 60 identical units only. This arrangement, containing one type of subunit environment, is called T=1. Complete viruses, such as southern bean mosaic virus, require larger shells (> 300Å). This virion is composed of 180 chemically identical subunits which are present as 60 copies of three distinct conformations, an arrangement known as T=3. An example of this arrangement is shown in Fig. 1B.

Almost all of the solved virus structures, including the small RNA viruses, have shell domains formed from an eight stranded β-barrel as shown in Fig. 2. However, this fold is not universal: The structure of the bacterial virus MS2 with ~ 300Å T=3 shell was solved recently, and shown to have a completely different fold (Valegård *et*

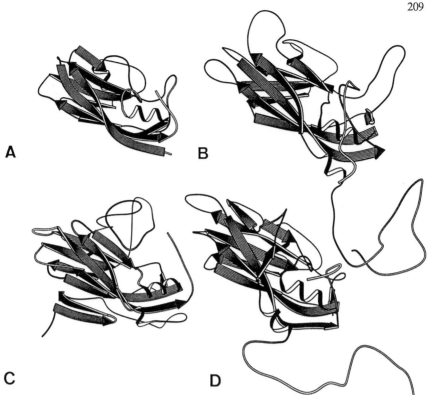

FIGURE 2. Structure of capsid proteins. (A) Shell domain of TBSV subunit in the A conformation, (B) rhinovirus VP1, (C) rhinovirus VP2, (D) rhinovirus VP3. The common 8 stranded
β-barrel motif should be apparent.

al., 1990). Adenovirus, a large DNA virus (1100Å), does form its shell domains from
a similar β-barrel-like fold, but uses it in a completely different way (Roberts and
Burnett, 1987).

The non-enveloped viruses which we will be using as models for the enveloped virus
cores, have T=3 or T=3-like shells composed of 180 units displaying the β-barrel fold.
The T=3 plant viruses have three conformations of their subunits. These are denoted
A, the subunit nearest the fivefold, and B and C which are adjacent to the threefold axis
(Fig. 3A). The A and B subunits are quite similar in conformation and in particular
exhibit virtually the same dihedral angle between adjacent units. The C subunits differ
in having a much flatter angle between units. The amino terminus of the C subunit fits
under the C–C dimer and widens the dimer angle. The complete T=3 structure can be
viewed as a dodecahedron of C subunits held together by amino-terminal arms with
the A and B subunits filling in the faces between the Cs. The amino-terminal arm has

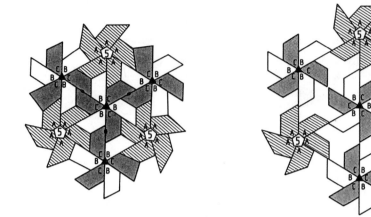

FIGURE 3. The surface of a T=3 shell in its normal (left) and expanded (right) organization. The expansion has caused the opening of the capsid in the regions corresponding to the divalent ion interaction sites in TBSV. Half of these sites correspond to the locations of the interaction with the spikes in the alphaviruses.

a very important role in this structure since it serves as a switch between the multiple conformations of the subunit required for the larger T=3 shell. The amino-terminal arm also has a role in the packaging of the viral genome since it interacts with it in the core. The A, B and C conformations form a quasi-equivalent trimer which is often held together by interaction with divalent ions. Removal of the divalent ions results in swelling of the capsid and the opening of holes between the subunits of this trimer (Fig. 3B).

The picornaviruses also have 180 subunits, but the three conformations are occupied by different polypeptide chains. The shell-forming domains of the three are very similar and their structural roles correspond to those of the T=3 virus so that VP1 acts as A, VP3 acts as B and VP2 acts as C. The elaborations which distinguish VP1, VP2, and VP3 occur as loops on the surface of the virion (Fig. 2). These loops contain the bulk of the antigenic determinants of the virus and most of the structural variation which occurs between the different picornavirus proteins. VP3 is the least elaborated of the three shell domains and, in that sense, is a prototype which is most easily related to the T=3 virus capsid proteins.

We know relatively little about the structures of enveloped viruses. The viruses which we will consider here, alphaviruses, hepadnaviruses and flaviviruses, are examples of the simplest type of enveloped virus. They form capsids from a single species of coat protein which package nucleic acid in the cytoplasm of animal cells. The alphaviruses, like Semliki Forest virus and Sindbis virus, bud from the plasma membrane. In the budding process, the capsids interact with the cytoplasmic tails of envelope glycoproteins excluding the host membrane proteins and deforming the membrane to envelope the capsid. The exclusion of host proteins is extremely efficient; none can be detected in the budded virion although the viral proteins comprise less than one percent of the plasma membrane proteins. Hepadnaviruses, such as hepatitis B, and flaviviruses, such as yellow fever virus, bud into early elements of the secretory pathway, but the topology of budding is the same. This interaction of the nucleocapsid with the spikes not only allows the virus to leave the cell without cell lysis but provides a way of altering the surface which the virus presents to the immune system of the host. The envelope proteins can be changed dramatically so that the epitopes present on the surface of the virus are altered while retaining the integrity of the capsid. In the picornaviruses, surface antigenicity is altered by mutations in the loop residues of the capsid proteins without altering the shell forming domains. The result in either case is that the virus cloaks itself from the immune system of the host while maintaining the structural framework needed for assembly.

3D Electron Microcopy of Enveloped Capsids

An understanding of the two stages of enveloped virus assembly, i.e. assembly of the nucleocapsid in the cytoplasm and interaction of the nucleocapsid with the envelope proteins to produce a budded virion, must be based on an understanding of the structures of the components involved. Unfortunately, obtaining high resolution structural information on enveloped viruses is hampered by the lack of three-dimensional crystals of viruses such as Semliki Forest or Sindbis suitable for X-ray crystallography. Our group and several others have produced three-dimensional crystals of alphaviruses; however, these crystals only appear to be ordered to approximately 40Å. In the absence of useful three-dimensional crystals, we are forced to use the second best method, electron microscopy. Conventional electron microscopy requires that the specimen be dried before being placed in the microscope vacuum and that it be stained to enhance its contrast. For an enveloped virus which has a lipid bilayer this is disastrous and results in images like that of Semliki Forest virus in Fig. 4A. An alternative method is cryo-electron microscopy, where drying is avoided by plunging a thin film of viruses in an aqueous buffer into ethane slush at near liquid nitrogen temperatures, cooling the

212

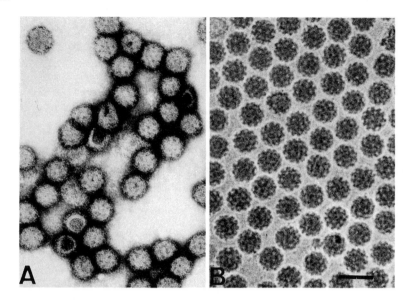

FIGURE 4. Electron micrographs of Semliki Forest virus negatively stained (A) and in vitreous water (B) show the better preservation of the structure after vitrification. Scale bar, 100 nm.

specimen so rapidly that it vitrifies. The vitrified state is preserved by maintaining the specimen at temperatures below $-160°C$ during transfer and subsequent observation in the microscope. The result is an excellent preservation of the structure (Fig. 4B) because its hydrated state has been maintained (Adrian *et al.*, 1984; Dubochet *et al.*, 1988). A further advantage is that the smooth water background allows specimen features to be visualized by phase constrast and obviates the need for contrast enhancing negative stains.

The electron micrograph of an unstained specimen in water represents its projected electron density. Combining the Fourier transforms of these projections yields a three-dimensional map of the electron density of the object. This process is very efficient for highly symmetric objects such as icosahedral viruses since the number of views required to define the density to a given resolution is fewer. Our group and others have developed algorithms for identifying the orientation of a symmetric object on the basis of symmetries in the Fourier transformed projection (Crowther, 1971; Fuller, 1987; Baker *et al.*, 1989). These computational tools allow one to obtain three dimensional structures from cryo-electron micrographs of unstained icosahedral viruses in a routine manner. Comparisons for specimens where both cryo-electron microscopic and X-ray diffraction data is available have shown very good agreement. Recently, our group in collaboration with Phoebe Stewart and Roger Burnett at the Wistar Institute in Philadelphia, have shown that the structure of adenovirus hexon seen in a icosahedral reconstruction from cryo-electron micrographs is the same as that calculated from the high resolution X-ray structure (Stewart, Burnett, Cyrklaff and Fuller, unpublished results). The combination of cryo-electron microscopy and image reconstruction allows reliable determination of icosahedral virus structure at moderate resolution without the need for crystallization.

A surface representation of the structure of Sindbis virus at 30Å resolution is shown in Fig. 5A (Fuller, 1987). The whole structure is about 700Å in diameter and includes spike glycoproteins which project $\sim$ 100Å from the membrane surface. The glycoproteins are clustered as trimers and arranged with triangulation number T=4. The signature of T=4 symmetry is the quasi-sixfold arrangement of spikes around the icosahedral two-fold axis. Hence, there are 240 copies (4 $\times$ 60) of each of the spike glycoproteins in the virion If the electron density corresponding to the membrane is removed to reveal the capsid (Fig. 5B), a very different arrangement of subunits is revealed, where the subunits are sixfold symmetrically arranged around the threefold axis, characteristic of T=3 symmetry and indicating that there are 180 subunits (3 $\times$ 60) in the capsid. Hence, during budding the 180 subunits of the T=3 symmetric capsid must organize 240 subunits of the T=4 envelope proteins. Analyses of similar reconstructions of capsids isolated by detergent treatment and of the closely related Semliki Forest virus, yield the same conclusions.

The observation of T=3 symmetry for the alphavirus nucleocapsid suggested that there may be something fundamentally similar between enveloped viral cores and the capsids of non-enveloped virus shells such as those of the picornaviruses described above. This idea gains further support from the correspondence between the positions of functional sites in the two types of systems. The interaction of the spikes with the nucleocapsid is the critical one which leads to virus budding. Examination of the arrangement of spikes and the nucleocapsid density reveals that the cytoplasmic tails of the spike proteins interact with the capsid at the positions of holes between the capsid subunits (Fig. 5B). The positions of three-quarters of these interaction sites (those nearest the five fold axes) correspond to the positions of the divalent ion sites which stabilize the subunit trimers in plant viruses such as tomato bushy stunt virus. The isolated nucleocapsid is less sturdy than the intact virion, suggesting that the interaction with spikes may also stabilize the nucleocapsid structure. The other quarter of the interaction sites are those on the threefold axes. Dr Catherine Vénien has completed a three-dimensional reconstruction which shows that treating Semliki

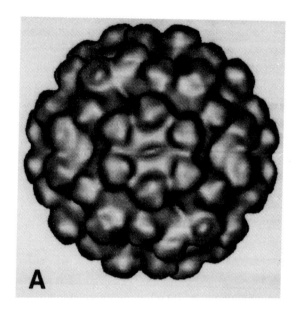

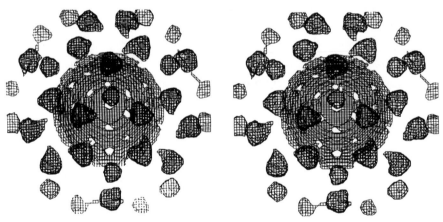

FIGURE 5. The three-dimensional reconstruction of Sindbis virus presented through an outer surface representation (A) revealing the trimeric spikes in a T=4 arrangement and in a stereo view with the membrane density suppressed (B) to show the positions of the spikes superimposed on capsid.

Forest virus capsids lightly with octyl- β-D-glucopyranoside selectively removes the threefold spikes, indicating that this interaction is weaker than the fivefold interaction corresponding to the divalent ion binding site (Vénien and Fuller, in preparation). The combination of the complementarity of the capsid interaction sites for envelope proteins and the possibility of weak interaction with a fraction of the spikes allows the combination of accuracy and flexibility required for budding. The coincidence of positions of the spike interaction and divalent ion interaction sites in enveloped and non-enveloped virus capsids is further support for their analogous roles in the structures.

Comparative Sequence Analysis

To explore the possibility that these structural homologies inferred from our 30Å resolution maps were supported by a corresponding homology between the sequences, we used a sensitive sequence comparison method developed by Dr Patrick Argos, in which sequences are compared by amino acid physical characteristics as well as by using the Dayhoff matrix, and allowing interactive picking of aligned sequence fragments (Argos, 1987). Several alphavirus sequences were compared against the sequences of several picornavirus VP3's, as well as with VP1 and VP2 for the same viruses, in a total of 30 pairwise comparisons. Although many of these comparisons did not show convincing homology (i.e. they did not contain unique diagonals with high correlation that could be followed throughout the β-barrel region of the picornavirus sequence), those that did shared homology over the same regions of the sequence. The best correspondence between an alphavirus nucleocapsid protein sequence and a picornavirus was between Semliki Forest virus and foot-and-mouth-disease virus VP3. Fig. 6A shows a homology plot for these two sequences. The aligment starts near SFV residue 95, which marks the end of the positively charged amino-terminal domain of the SFV capsid protein. By analogy with the T=3 plant viruses, this region would interact with nucleic acid within the nucleocapsid and act as the A,B vs C switch in conformation. In the middle region, corresponding to β strands C, D and E, there is some ambiguity in the assignment, where it cannot be clearly deduced from the homology plots. The alignment appears much more significant than the alignment between, for example, VP2 and VP3 from two picornaviruses (Fig. 6B), corresponding to a known structural homology. Further, comparisons between a picornavirus or an alphavirus and MS2 (Fig. 6C) yields homology plots with no detectable homology above the noise level. This is reassuring since crystallography has shown this capsid protein to consist of a completely different fold.

Despite these encouraging results, this sequence comparison approach is flawed both by its lack of ability to detect true structural homology and the occurrence of false positives. As long as the comparison of pairs of sequences is the basis for the alignment, one can argue that a capsid protein which does not match one tested sequence will match well with another sequence. This means that the lack of homology may reflect the choice of partners in the comparison rather than the absence of the fold in the capsid being examined. By using a search probe of several sequences which are aligned themselves on the basis of their known structural homology, and comparing this set of sequences with a test sequence, we hoped to increase the sensitivity and reliability of this approach.

One probe consisted of a set of four aligned picornaviral VP3's (mengo, polio, FMD and rhino; Luo *et al.*, 1987; Acharya *et al.*, 1989), with a relatively high degree of sequence and structural homology. A similar probe was made from the structurally aligned sequences of VP1, VP2 and VP3 from rhino and mengo virus (Luo *et al.*, 1987) which display much weaker sequence homology.

Fig. 7A shows the homology plot for the comparison of the VP3 probe and SFV. The aligment is in agreement with the pairwise comparisons. Comparing the whole set of alphaviruses simultaneously to the VP3 probe (Fig. 7B) yields a similar aligment, and further supports our assignment for the middle region of the sequence.

The VP1-VP2-VP3 probe produced no convincing alignments of the picorna and alphavirus sequences. This probe was not more sensitive for detecting the known structural homology between picornaviral VP's. Only relatively strong homologies to the sequences contained in the probe could be detected. The plots are noisy and difficult to interpret for more distantly related sequences. We conclude that multiple sequences aligned by known structural homology serves as a better probe only when the sequences also share good sequence homology.

Fig. 8 shows the resulting alignment of the alphavirus sequences with those of several picornaviruses based on the pairwise and multiple sequence comparisons. Some picornavirus sequences are aligned so that the known structural homology is imposed; additional sequences were included by sequence homology. The combination of pairwise and multiple comparisons allows us to assign the alignments confidently over most of the sequence; however, there is always some ambiguity in such an alignment, and its reliability must be assessed through experimental test.

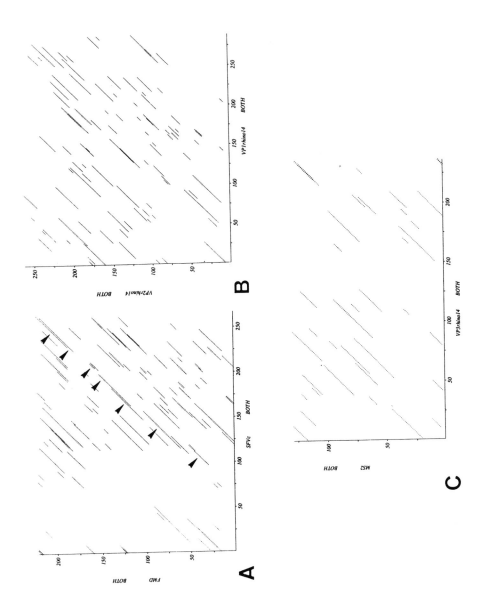

FIGURE 6. Homology plot of pairwise comparisons: (A) SFV versus FMD VP3 (B) rhino VP2 versus rhino VP3 (C) rhino VP3 versus MS2. Type of line shows number of standard deviations (SD) above mean: ------ $3.3 \leq SD < 4.0$; IIIIII $4.0 \leq SD < 4.5$; IIIIII $SD \geq 4.5$. Arrows indicate aligned sequence fragments.

218

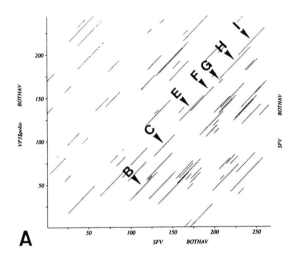

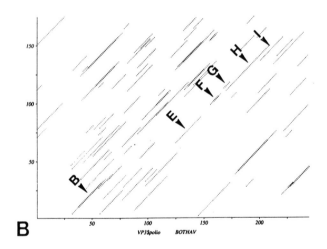

FIGURE 7. Multiple comparisons: (A) SFV versus multiple picornavirus VP3 probe. ------ 3.0 ≤ SD < 3.3; ⅲⅲⅲ 3.3 ≤ SD < 4.0; ▮▮▮▮ SD ≥ 4.0. (B) Multiple alphavirus versus multiple picornavirus VP3 probe. ------ 2.7 ≤ SD < 3.0; ⅲⅲⅲ 3.0 ≤ SD < 3.5; ▮▮▮▮ SD ≥ 3.5. Note that the cutoff level for multiple comparisons can be lower that for the pairwise comparisons because an average of all pairwise comparisons is taken at each point. Arrows indicate aligned sequence fragments with the corresponding secondary structure element marked.

Structural Model and Experimental Tests

One way to evaluate the significance of this alignment is to make a model of the alphavirus structure based on the alignment and the known fold of the picornavirus proteins. This model can then be tested experimentally by examining the predicted features of exposure and juxtaposition of residues (Fuller and Argos, 1987). For the alphavirus case, we generated a model by folding the SFV sequence into the form of the rhinovirus VP3 (Fig. 13A). Such a model must be seen only as an indication of the putative fold, and the details of the structure, particularly in the loop regions, remain undetermined. The SFV capsid protein has a serine protease-like autoproteolytic activity which is responsible for its cleavage from the polyprotein precursor which also contains the spike glycoprotein. Mutagenesis experiments have shown that this activity requires three residues, His-145, Asp-151 and Ser-21, which form the catalytic site. Although these residues are separated in the sequence, they are juxtaposed in our folding model as required by their contribution to the active site. A second test is the exposure of residues to lactoperoxidase labelling in the intact capsid. Coombs and Brown, (1987) have shown that while all four tyrosines (Tyr's 162, 180, 189 and 198) in the Sindbis capsid protein can be labelled in denatured capsid protein, only one (Tyr 180) can be labelled in the intact capsid. This tyrosine is the only one of the four which is predicted to be at the capsid surface by the folding model.

This consistency of the folding model with experimental evidence is reassuring; however, one would like to go further and use the model to predict the effect of mutations on folding and assembly. This requires the establishment of a convenient expression and assembly system. This was not available for the alphaviruses, but it was well established for the hepadnaviruses. The hepadnaviruses are biologically very different from the alphaviruses: although the first step in assembly involves the packaging of an RNA into the capsid, the mature virion carries a DNA genome and a reverse transcriptase rather than an RNA genome like that of the picornaviruses or the alphaviruses.

Despite this fundamental difference, the hepadnavirus core protein shows quite significant homology to the picornavirus VP3 proteins. The homology of the hepadnavirus core protein to the mengo virus VP3 was more apparent than was the alphavirus case (Argos and Fuller, 1988). The homology once again extends through the β-strand regions of the picornavirus structures but differs from the alphavirus alignment in that the hepadnavirus amino terminus is homogous to the beginning of the β-barrel while the homology is lost toward the positively charged carboxy terminus. The homology is seen in all the members of the hepadnavirus family including the related viruses in woodchucks, ground squirrels, ducks and herons.

We applied the same method to generate a model from the aligned sequences using the known fold of the mengo virus and the sequence of hepatitis B virus (Fig. 13B; Argos and Fuller, 1988). The location of the dominant antigenic sites on the capsid provided a first test of the model. Anti-peptide antibodies have been raised against a large number of sequences from the core protein. Two sequences, marked e1 and e2, give rise to antibodies which show selective reactions with intact versus denatured core protein (Salfeld, 1985). Antibodies against e1 react with intact cores while those against e2 react only with the unfolded protein. The folding model explains this behavior. The e1 sequence is in a region corresponding to the picornavirus "puff" and should be exposed on the capsid surface while e2 would be buried at a subunit interface and inaccessible in intact capsid.

Expression of hepatitis virus capsid protein in E. coli leads to formation of cores which package RNA. This provided the convenient expression system needed to test predictions about the structure based on the picornavirus model. Cryo-electron microscopy and icosahedral image reconstruction of these E. coli-produced cores established that the hepatitis core displayed the expected T=3 symmetry. This reconstruction also revealed that the RNA was located predominantly against the interior wall of the core. This arrangement may be significant in vivo since the virus contains a polymerase which could fill the space left vacant by the RNA (von Bonsdorff, Nassel, Cyrklaff and Fuller, unpublished results).

Using this expression system, we examined the effects of two types of mutations. The first, and crudest, were mutations that would be expected to disrupt the fold and destroy the ability of the core to assemble. Many such mutations have been tried and,

---

[1]First line shows the structural elements in the picornaviruses. Uppercase letters represent β-strand, lowercase letters α-helix structural elements, Nomenclature after Harrison et al. (1978). The last line shows the concensus sequence. Symbols (#, @, %, *, o) indicate that residue is conserved in nearly all sequences [Conservation groups: #=(AVILFM), @=(FYHW), %=(EDQN), *=(PG), o=(ST)]. Capital letters indicate that a residue is strictly conserved in all of the sequences, small letters that a residue is present in more than two sequences in one set and one in the other. Virus abbreviations: Mengo – monkey mengoencephalomyocarditis virus (Luo et al., 1987); Tmev – Theiler's murine encephalomyelitis virus (Pevear et al., 1987); FMD – foot-and-mouth virus A (Carroll et al., 1984); rhino – human rhinovirus 14 (Stanway et al., 1984); polio – poliovirus type 1 Sabin (Nomoto et al., 1982); Cox – Coxsackievirus B1 (Iizuka et al., 1987); SFV – Semliki Forest Virus (Garoff et al., 1980); rrv – Ross River virus (Faragher et al., 1988); SNV – Sindbis virus (Strauss et al., 1984); vee – eastern equine encephalomyelitis virus (Chang and Trent, 1987); veev – Venezuelan equine encephalitis virus (Johnson et al., 1986).

```
1    --------------------------------------------zzzzzzz-  struct
1    SPIPVTIREHAGTWYSTLPDSTVPIYGKTPVAPANYMVGEYKDFLEIAQI    Mengo
1    SPIPVTVREHKGCFYSTNPDTTVPIYGKTISTPSDYMCGEFSDLLELCKL    Tmev
1    GIFPVACADGYGGLVTTDPKTADPVYGKVYNPPKTNYPGRFTNLLDVAEA    FMD
1    -GLPTTTLPGSGQFLTTDDRQSPSALPNYEPTPRIHIPGKVHNLLEIIQV    rhino
1    -GLPVMNTPGSNQYLTADNFQSPCALPEFDVTPPIDIPGEVKNMMELAEI    polio
1    -GLPVMTTPGSTQFLTSDDFQSPSAMPQFDVTPEMQIPGRVNNLMEIAEV    Cox
1    -------------------------DKQADKKKKKPGKRERMCMKIEN      SFV
1    -------------------------KPKPQAKKKKPGRRERMCMKIEN      rrv
1    -------------------------KKKKQPAKPKPGKRQRMALKLEA      SNV
1    -------------------------PAKKQKRKPKPGKRQRMCMKLES      vee
1    -------------------------NGNKKKTNKKPGKRQRMVMKLES      veev
1    -------------------------n----pk---pGk---#---#%a      concensus


8    -BBBB---------------BBBBB--------CCCC------------a    struct
51   PTFIGNK-MPN---A--VPYIEASN-TAVKTQPLAVYQVTLSCSCLAN-T    Mengo
51   PTFLGNPNTNN----KRYPYFSATN-SVPAT-SMVDYQVALSCSCMAN-S    Tmev
51   CPTFLRF-D-D---G--KPYVVTRA-D-D-TRLLAKFDVSLAAKHMSN-T    FMD
50   DTLIPMNNT-HTKDEVNSYLIPLNANRQNE--QVFGTNLFIGDGVFKT-T    rhino
50   DTMIPFDLSAKKKNTMEMYRVRLSDKPHTDD-PILCLSLSPASDPRLSHT    polio
50   DSVVPVNNTDNNVNGLKAYQIPVQSNSDNR-RQVFGFPLQPGANNVLNRT    Cox
24   DCIFEVKH-E-------------------GKVTGYACLVGDKVMKP-A     SFV
24   DCIFEVKL-D-------------------GKVTGYACLVGDKVMKP-A     rrv
24   DRLFDVKN-ED------------------GDVIGHALAMEGKVMKP-L     SNV
24   DKTFPIML-N-------------------GQVNGYACVVGGRVFKP-L     vee
24   DKTFPIML-E-------------------GKINGYACVVGGKLFRP-M     veev
11   d-t#*#k--%------------------#-g@--##g--v#k---        concensus


22   aaaaaaaaaDDDDDDDDDDDDDDD----EEEEEEEEE---------bbb-    struct
93   FLAALSRNFAQYRGSLVYTFVFTGTAMMKGKFLIAYTPPGAG-KPTSRD-   Mengo
94   MLAAVARNFNQYRGSLNFLFVFTGAAMVKGKFLIAYTPPGAG-KPTTRD-   Tmev
90   YLSGIAQYYTQYSGTINLHFMFTGSTDSKARYMVAYIPPGVETPPDTPE-   FMD
96   LLGEIVQYYTHWSGSLRFSLMYTGPALSSAKLILAYTPPGAR-GPQDRR-   rhino
99   MLGEILNYYTHWAGSLKFTFLFCGSMMATGKLLVSYAPPGAD-PPKKRK-   polio
99   LLGEILNYYTHWSGSIKLTFMFCGSAMATGKFLLAYSPPGAG-VPKNRR-   Cox
51   HVKGVIDNADLAKLAF----------KKSSKYDLECAQIPVHMRSDASKY   SFV
51   HVKGTIDNPDLAKLTY----------KKSSKYDLECAQIPVHMKSDASKY   rrv
52   HVKGTIDHPVLSKLKF----------TKSSAYDMEFAQLPVNMRSEAFTY   SNV
51   HVEGRIDNEQLAAIKL----------KKASIYDLEYGDVPQCMKSDTLQY   vee
51   HVEGKIDNDVLAALKT----------KKASKYDLEYADVPQNMRADTFKY   veev
27   -#-gv#%n-------#------------s-k@-#-@a--*#--k-dt-k-   concensus


58   -bbbbbFFFFFFFF-----GGGGGGG--------GGG-----------HHH  struct
141  -QAMQATYAIWDLG-LN-SSYSFTVPFISPTHFRMVGTDQANITNVDGWV  Mengo
142  -QAMQSTYAIWDLG-LN-SSFNFTAPFISPTHYRQTSYTSPTITSVDGWV  Tmev
139  -EAAHCIHAEWDTG-LN-SKFTFSIPYVSAADYAYTASDTAETTNVQGWV  FMD
144  -EAMLGTHVVWDIG-LQ-STIVMTIPWTSGVQFRYTDPD--T-YTSAGFL  rhino
147  -EAMLGTHVIWDIG-LQ-SSCTMVVPWISNTTYRQTIDD--S-FTEGGYI  polio
147  -DAMLGTHVIWDVG-LQ-SSCVLCVPWISQTHYRYVVED--D-YTAAGYV  Cox
91   THEKPEGHYNWHHGAVQYSGGRFTIP----TGAGKPGDSGRPIFDNKGRV  SFV
91   THEKPEGHYNWHHGAVQYSXGRFTIP----TGAGKPGDSGRPIFDNKGRV  rrv
92   TSEHPEGFYNWHHGAVQYSGGRFTIP----RGVGGRGDSGRPIMDNSGRV  SNV
91   TSDKPPGFYNWHHGAVQYENNRFTVP----RGVGGKGDSGRPILDNKGRV  vee
91   THEKPQGYYSWHHGAVQYENGRFTVP----KGVGAKGDSGRPILDNQGRV  veev
46   -------@--W--G-#Q-s---ft#P-----------------f---G-#  concensus


83   HHHHHHHH-------IIIIIIIIIIIIIIIIIIII----------------  struct
188  TVWQLTPLTYPPGCPTSAKILTMVSAGKDFSLKMPISPAPWSPQ------  Mengo
189  TVWKLTPLTYPSGTPTNSDILTLVSAGDDFTLRMPISPTKWVPQ------  Tmev
186  CVYQIT-HGK----AENDTLLVSASAGKDFELRLPIDPRTQ---------  FMD
188  SCWYQTSLILPPETTGQVYLLSFISACPDFKLRLMKDTQTISQTVALTE-  rhino
191  SVFYQTRIVVPLSTPREMDILGFVSACNDFSVRLMRDTTHIEQK-ALAQ-  polio
191  TCWYQTNIIVPADVQSTCDILCFVSACNDFSVRMLKDTPFIRQD-NFYQ-  Cox
137  VA-----IVLGGANEGSRTALSVVTWN-KDMVTRVTPEGSEEW-------  SFV
137  VA-----IVLGGANEGARTALSVVTWT-KDMVTRVTPEGTEEW-------  rrv
138  VA-----IVLGGADEGTRTALSVVTWNSKGKTIKTTPEGTEEW-------  SNV
137  VA-----IVQGGVNEGSRTALSVVTWNQKGVTVKDTPEGSEPW-------  vee
137  VA-----IVLGGVNEGSRTALSVVMWNEKGVTVKYTPENCEQW-------  veev
59   -#-----###**---gs-t#Ls##o------v-----------------  concensus
```

FIGURE 8. Alignment of all picornavirus sequences with several alphavirus sequences[1].

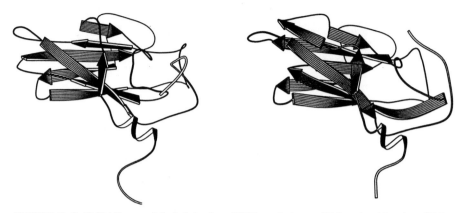

FIGURE 9. (left) Folding model of alphavirus (SFV) made by modifying the rhinovirus VP3 fold, allowing for deletions and insertions according to the alignment shown in Fig 8. (right) Folding model of Hepatitis B, made in the same way, using mengovirus VP3.

in brief, any mutation which should cause a significant disruption in the predicted β-strands results in no assembled capsids. The one region in the body of the sequence which can be changed without affecting capsid formation is that corresponding to the "puff" or e1 region. Fairly large insertions can be placed at this site; this has led to proposals that the assembled capsid would serve as a good carrier for foreign epitopes which could provide a basis for vaccines. The second and more subtle type of mutation is one which should allow assembly but change the properties of the assembled core in a predictable way. In the T=3 plant viruses, the long, positively charged amino-terminal region has the dual function of forcing the switch between the conformations of the capsid protein needed to form the T=3 shell and interacting with the nucleic acid within the capsid. The hepadnavirus amino-terminal region is short, and we asked whether the longer carboxy-terminal tail of the hepadnavirus had taken over these functions. A series of deletions were made from the carboxy-terminal tail and the effect on assembly into capsids and packaging of nucleic acid was followed. Cryo-electron microscopy showed that the mutants with truncated tails could assemble, although the capsids were progressively less stable as the deletions brought the carboxy-terminus closer to the β-barrel. The assembled capsids were the same diameter as those of the full length wild type sequence, showing that the carboxy terminal tail is not the switch which allows the capsid protein to take on the three conformations necessary to make a T=3 capsid. Examination of the reconstructions showed that the layer of RNA seen against the wall of the wild type capsid is not present in the capsids formed from the deleted protein

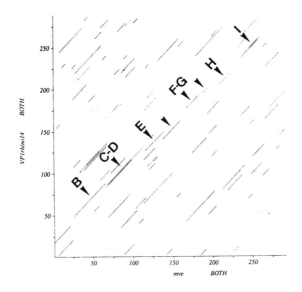

FIGURE 10. Homology plot of comparison between the flavivirus Murray Valley encephalitis virus (mve; Dalgano *et al.*, 1986) and rhinovirus VP1. Aligned fragments and corresponding secondary structure elements in the VP1 protein indicated by arrows. ------ $3.4 \le SD < 3.8$; ııııı $3.8 \le SD < 4.5$; ııııı $SD \ge 4.5$.

(von Bonsdorff, Nassel, Cyrklaff and Fuller, in preparation). We conclude that only the packaging function of the aminoterminal tail has been transferred to the hepadnavirus carboxy terminus.

Conclusions

We have shown that the picornavirus capsid folds can be used as models for two families of enveloped virus capsids. Argos, (1989) has made a similar argument for retrovirus p24 proteins and advanced a model for the human immunodeficiency virus protein. A natural question is whether all the enveloped virus capsids follow this folding scheme. As a first step, we asked whether we could use these alignment procedures to include the flaviviruses in the group of β-barrel type viruses, alternatively confidently exclude them. The flaviviruses are superficially similar to the alphaviruses, with which they were formerly included, and are believed to contain an icosahedral capsid. Fig. 10 shows a homology plot for a comparison between a flavivirus sequence (Murray Valley encephalitis virus) and rhinovirus VP1. There are diagonals showing high homology in the beginning of the sequence. This homology is apparently maintained throughout the VP1 sequence; regions corresponding to β-strand elements are the most conserved. It can be seen from this alignment that some of the conserved cysteines in the flavivirus sequence could be used to stabilize a loop between the G and H β-strands, but further data are needed to evaluate this model. This approach to structure prediction is still most valuable as a motivation for precise experimental tests.

References

Acharya, R., Fry, E., Stuart, D., Fox, G., D., R., and Brown, F. (1989). *Nature*, 337:709–716.

Adrian, M., Dubochet, J., Lepault, J., and McDowall, A. W. (1984). *Nature*, 308:32–36.

Argos, P. (1987). *J. Mol. Biol.*, 193:385–396.

Argos, P. (1989). *EMBO J.*, 8:779–785.

Argos, P. and Fuller, S. D. (1988). *EMBO J.*, 7:819–824.

Baker, T. S., Drak, J., and Bina, M. (1989). *Biophys. J.*, 55:243–253.

Carroll, A. R., Rowlands, D. J., and Clarke, B. E. (1984). *Nucleic Acids Res.*, 12:2461–2472.

Chang, G. J. J. and Trent, D. W. (1987). *J. Gen. Virol.*, 68:2129–2142.

Coombs, K. and Brown, D. T. (1987). *Virus Res.*, 7:131–149.

Crowther, R. A. (1971). *Phil. Trans. Roy. Soc. London*, B261:221–230.

Dalgano, L., Trent, D. W., Strauss, J. H., and Rice, C. M. (1986). *J. Mol. Biol.*, 187:309–323.

Dubochet, J., Adrian, M., Chang, J.-J., Homo, J.-C., Lepault, J., McDowall, A. W., and Schultz, P. (1988). *Quart. Rev. Phys.*, 21:129–228.

Faragher, S. G., Meek, A. D. J., Rice, C. M., and Dalgarno, L. (1988). *Virology*, 163:509–526.

Fuller, S. D. (1987). *Cell*, 48:923–934.

Fuller, S. D. and Argos, P. (1987). *EMBO J.*, 6:1099–1055.

Garoff, H., Frischauf, A.-M., Simons, K., Lehrach, H., and Delius, H. (1980). *Proc. Nat. Ac. Sci. USA*, 77:6376–6380.

Harrison, S. C., Olson, A. J., Schutt, C. E., Winkler, F. K., and Bricogne, G. (1978). *Nature*, 276:368–373.

Hogle, J. M., Chow, M., and Filman, D. J. (1985). *Science*, 229:1358–1365.

Iizuka, N., Kuge, S., and Nomoto, A. (1987). *Virology*, 156:64–73.

Johnson, B. J. B. and Kirmey, R. M. Kost, C. L. (1986). *J. Gen. Virol.*, 67:1951–1960.

Luo, M., Vriend, G., Kramer, G., Minor, I., Arnold, E., Rossmann, M. G., Boege, U., Scraba, D. G., Duke, G. M., and Palmenberg, A. C. (1987). *Science*, 235:182–191.

Nomoto, A., Omata, T., Toyoda, H., Kuge, S., Horie, H., Kataoka, Y., Genba, Y., Nakano, Y., and Imura, N. (1982). *Proc. Nat. Acad. Sci. USA*, 79:5793–5797.

Pevear, D. C., Canenoff, M., Rozhon, E., and Lipton, H. L. (1987). *J. Virol.*, 61:1507–1516.

Roberts, M. M. and Burnett, R. M. (1987). In Burnett, R. and Vogel, H., editors, *Biological Organization: Macromolecular Organization at High Resolution*, pages 113–124. Academic Press.

Rossmann, M. G., Arnold, E., Ericksson, J. W., Frankenberger, E. A., Griffith, J. P., Hecht, H.-J., Johnson, J. E., Kamer, G., Luo, M., Mosser, A. G., Rueckert, R. R., Sherry, B., and Vriend, G. (1985). *Nature*, 317:145–153.

Salfeld, J. (1985). PhD thesis, Rupert Karls Universität, Heidelberg.

Stanway, G., Hughes, P. J., Mountford, R. C., Minor, P. D., and Almond, J. W. (1984). *Nucleic Acids Res.*, 12:7859–7875.

Strauss, E. G., Rice, C. M., and Strauss, J. H. (1984). *Virology*, 133:92–110.

Valegård, K., Liljas, L., Fridborg, K., and Unge, T. (1990). *Nature*, 345:36–41.

Varghese, J. N., Laver, W. G., and Colman, P. M. (1983). *Nature*, 303:35–40.

Wilson, I., Skehel, J., and Wiley, D. (1981). *Nature*, 289:366–373.

Discussion

Q: Are there hepatitis virus capsid protein sequences which are sufficiently distant from the human virus sequence that they would serve as a test of the folding models? For example is there a large insertion or deletion in one of the sequences?

A: You're thinking of the sequence of duck hepatitis virus. The sequences of all the human hepatitis B virus capsids are very closely related as are the sequences of the other mammalian hepadnaviruses. The largest change found in the avian hepadnaviruses, where an insertion of 38 amino acids is seen at position 131 through 169 relative to the aligned mammalian virus sequences. The duck virus also has a different morphology from the mammalian hepatitis viruses. Its capsid appears more spikey, consistent with the projection of an extra domain of protein from the capsid surface. This insertion is easily accommodated by the model since it occurs at the position of the e1 epitope. The duck insertion can be exchanged with the corresponding sequences in the human capsid sequence and a properly assembled particle is produced. Hence the observations of altered antigenicity between the duck and the human virus capsids, the relatively rougher appearance of the avian capsid and the position of the insertion within a surface loop of the model are good support for the folding model. The other mammalian viruses are so similar that they provide relatively poor tests of the model and for this reason we turned to site directed mutagenesis.

Q: Would it not be possible to use cryo-electron microscopy with the small crystals of Semliki Forest virus which you have already obtained? That would give the advantages of increased signal to noise through averaging the units of the crystal while allowing you to use your present algorithms for reconstruction and the low temperature techniques for preserving the specimen

A: Unfortunately, the crystals are too small for X-ray work and too large for electron microscopy. Electrons interact more strongly with matter than X-rays and so undergo multiple scatters when passing through an object that is even a few thousand angstroms thick. Once the thickness gets above 100 microns as in these crystals, the information conveyed is reduced by multiple scattering to extremely low resolution. It would be necessary to work with thinner crystals. Unfortunately these virus crystals are not terribly stable as single layers. We're attempting to form thin sheets by crystallisation on lipid films but that has yet to yield highly ordered arrays. At present, the Semliki Forest virus 3D crystals diffract X-rays to about 40Å resolution while the reconstruction from cryoelectron micrographs has a resolution of 30Å. Cryo-electron microscopy is actually more successful for this problem than X-ray diffraction at the moment. We hope to obtain better crystals or, failing that, increase the resolution of the information obtained by the image reconstruction. In theory, a high resolution reconstruction is possible from cryo-electron micrographs as shown by the work on crystalline specimens such as bacteriorhodopsin.

Structural Motifs of the Extracellular Matrix Proteins Laminin and Tenascin

Konrad Beck[1], Jürg Spring[2], Ruth Chiquet-Ehrismann[2],
Jürgen Engel[3] and Matthias Chiquet[3]

Institute for Biophysics[1]
Johannes-Kepler-University
A-4040 Linz-Auhof
Austria

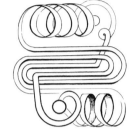

Introduction

Laminin and tenascin are two major extracellular matrix glycoproteins. They both consist of large disulphide-linked subunits composed of multiple structural and functional domains which are reflected in a distinct pattern of sequence motifs. These molecules belong to different protein families for which more and more members are being discovered. Members of these families have been discovered down to the level of *Anthomedusae* laminin (cf. Beck *et al.*, 1990) and leech tenascin (Masuda-Nakagawa *et al.*, 1989). The molecular structure not only varies considerably between species but for laminin also differences depending on the state of development and tissue origin have been elucidated. Varying numbers of tenascin isoforms generated by alternative splicing are found during development and in different tissues.

Laminin is predominently located in basement membranes where it is the most abundant noncollagenous protein. Due to its capability of self-association to large aggregates and the binding capacity for several other molecules, laminin is thought to play a crucial role in the structural organization of basement membranes. During embryogenesis it appears among the first extracellular components before collagens and nidogen. Biological activities attributed to laminin include the promotion of cell adhesion and migration, polarity, differentiation, neurite outgrowth and tumour metastasis. By means of antibodies, proteolytic fragments and synthetic peptides some of these functions have been correlated to specific structural elements (for reviews see Engel and Furthmayr, 1987; Ekblom, 1989; Timpl, 1989; Beck *et al.*, 1990; Timpl *et*

[2]Friedrich-Miescher-Institut, P.O.B. 2543, CH-4002 Basel, Switzerland
[3]Department of Biophysical Chemistry, Biocenter of the University, CH-4056 Basel, Switzerland

al., 1990).

Tenascin has been discovered independently by several laboratories and thus several synonyms are used (cytotactin; GMEM: glioma mesenchymal extracellular matrix antigen; hexabrachion, J1 glycoprotein; myotendinous antigen). During development, wound healing and in tumours it is expressed transiently whereas in adult tissues it is found permanently at several locations thought to have a low turnover including gizzard, skin, brain, kidney and myotendinous junctions. Based on *in vitro* experiments several biological activities are proposed including the promotion of cell attachment (not spreading), cell growth, neurite outgrowth and hemagglutination (for reviews see Erickson and Lightner, 1988; Erickson and Bourdon, 1989; Chiquet 1989; Chiquet-Ehrismann, 1991).

The present work focusses on the analysis of the amino acid sequences of laminin and tenascin with respect to their organization into distinct domains. These can be correlated to specific structural elements visualized on electron micrographs. Finally a strategy for the analysis of large extended proteins is proposed which might be helpful for the mapping of functional epitopes to structural segments.

Structural Organization of Laminin and Tenascin

Most biochemical and structural investigations have been performed on laminin purified from a mouse tumour. The native molecular mass is about 10^6. After reduction it is dissociated into one heavy chain A ($M_r \sim 440,000$) and two different light chains B1 and B2 ($M_r \sim 220,000$). When examined by electron microscopy laminin reveals the shape of a cross (Fig. 1a) with three short and one 77 nm long arm (Engel *et al.*, 1981). Two of the 34 nm short arms contain a central and a terminal globule whereas the somewhat extended arm (48 nm) contains an additional one near the center of the cross (Bruch *et al.*, 1989). The long arm is terminated by two closely spaced globules. After negative staining three subdomains can be recognized in the globule adjacent to the 3 nm thick rod and two subdomains in the distal one (Bruch *et al.*, 1989; Beck *et al.*, 1990). By limited proteolysis with different proteases laminin can be dissected into several fragments which can be related to the entire molecule (Fig. 1b–g; for details see figure legend).

For chicken tenascin purified from embryo fibroblast conditioned medium molecular masses of about 1.2×10^6 and 600,000 have been determined. After reduction both species of molecules dissociate into monomers of $M_r \sim 190,000, 200,000$ and

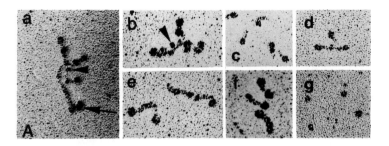

FIGURE 1. Electron micrographs of mouse tumour laminin and fragments thereof after rotary shadowing (bar: 100 nm)

Laminin (a) consists of three short arms one of which contains three instead of two globular domains (arrowhead). The long arm is terminated by two closely spaced globules (arrow). With cathepsin G laminin can be dissected into two large fragments, C1-4 (b) which consists of the short arms and C8-9 (e) comprising nearly the entire long arm without the outer globule. With elastase fragment E1-4 similar to C1-4 can be generated which can be cleaved to E4 (c) containing the amino-terminal domain of the B1-chain and a short rod. From the long arm fragments E8 (f) and E3 (g) arise which represent the distal half of the long arm and the outer globular domain, respectively. Pepsin treatment results mainly in fragment P1 (d) which resembles the central rod-like portions of the short arms.

230,000 (see e.g. Spring et $al.$, 1989). When purified from chicken gizzard the molecular mass of 200,000 is the same both under reducing or nonreducing conditions (Chiquet et $al.$, 1991). On electron micrographs this molecule exhibits a rod-like shape with a thin and a somewhat thicker half terminated by a globular domain (Fig. 2a). The most prominent shape of tenascin purified from medium is shown in Fig. 2c. It appears as a hexameric structure. Two sets of three arms are connected in a short rod to the central globule on opposite sites. The proximal third of the about 75 nm long arms appears significantly thinner than the distal two thirds. All rods end up in a globular domain. Beside these hexabrachions also half-hexabrachions and (rarely) non-abrachions (Fig. 2b and d) can be found. Each arm of the oligomeric structures corresponds to one polypeptide chain which is disulphide-linked within and probably near to the central domain.

The localization of different structural and functional sites has been elucidated by antibodies and fragments. A direct mapping of some epitopes has been performed by electron microscopy (Fig. 2e–g). The structure of some recombinant as well as proteolytic fragments is shown in Fig. 2. Limited digestion of tenascin with pepsin results in fragments which correspond to the thin proximal rods (Fig. 2k). After treatment with pronase parts of the distal rod and the terminal globule remain intact (Fig. 2l). A similar fragment is generated by trypsin in the presence of Ca^{++} and

232

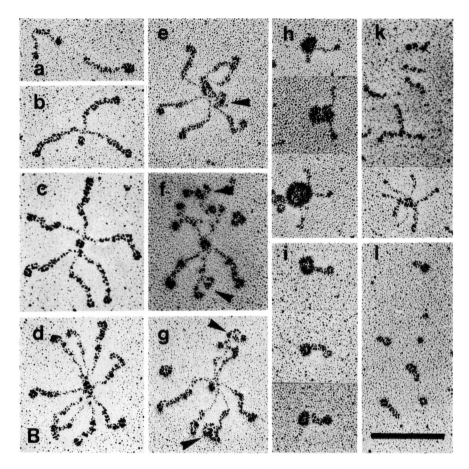

FIGURE 2. Electron micrographs of chicken tenascin and fragments thereof after rotary shadowing (bar: 100 nm)

Tenascin monomers assemble into various multimeric structures. From chick gizzard tenascin can be purified as a monomer (a) appearing as a linear array of thin and thicker rods with a terminal globule. Tenascin from fibroblast conditioned medium results in tri- (b), hexa- (c; major fraction) and nonabrachionic (d) aggregates. Also oligomers with 12 arms are reported (Erickson and Lightner, 1988). Distinct epitopes of tenascin are localized by monoclonal antibodies: anti-Tn60 binds to the central globule (e), anti-Tn26 binds to the extra domains of Tn230 within the thick distal rod (f) and the epitop for anti-Tn68 which inhibits cell binding is localized near to the distal globule. The shape of β-galactosidase-tenascin fusion proteins appears similar to the corresponding parts of mature tenascin: λcTn8 (h) resembles the inner thin and λcTn200 (i) the outer thick rod domain including the terminal globule. Pepsin (k) and pronase (l) generate fragments corresponding to the distal and proximal domains, respectively.

specifically originates from the 230 kD variant (Chiquet *et al.*, 1991). Related fragments can be prepared by chymotrypsin (CT) and cyanogen bromide (CNBr) (Friedlander *et al.*, 1988) for which the N-termini have been determined allowing an exact correlation within the entire cytotactin/tenascin structure (Jones *et al.*, 1989; see Fig. 4). Antibodies against the distal 35k CNBr fragment inhibit the binding of cells, fibronectin and a cytotactin binding proteoglycan (Friedlander *et al.*, 1988).

Sequence Arrangement of Laminin and Tenascin

During the last years the complete cDNA derived protein sequences of several laminin chains have become available (cf. Hunter *et al.*, 1989; Timpl, 1989) but presently only for mouse both the hetro B-chains (B1 and B2) and an A-chain, each of them encoded by a separate gene, are sequenced (Sasaki *et al.*, 1988). The number of residues and the corresponding protein molecular masses are 1765 (194.6 kD; B1), 1574 (173.9 kD; B2) and 3060 (335.7 kD; A). The difference to the experimentally determined M_r is mainly due to asparagine-linked carbohydrates. Whereas the B-chains contain 14 putative acceptor sites (NxS or NxT) the A-chain has 44. All three chains are homologous to each other with the exception of the C-terminal domain of the A-chain which has no counterpart within the B-chains.

Each of the short arms is formed by the amino-terminal regions of the three chains (1158 residues of B1, 994 residues of B2 and 1537 residues of A). The following 570 residues of each chain form together the long arm. Its globular extension is formed by the 950 C-terminal residues of the A-chain. An overview on the sequence arrangment is depicted in Fig. 3 which will be discussed in detail later.

The complete sequences for chicken tenascin or cytotactin, respectively, are available from two laboratories (Jones *et al.*, 1989; Spring *et al.*, 1989). Unfortunately they differ in several positions. Here we will refer to the version and numbering of Spring *et al.* The overlapping cDNA clones contain the information for three alternative splicing variants with open reading frames coding for 1808 (198.9 kD), 1626 (179.0 kD) and 1535 (168.9 kD) residues. These variants correspond to the three monomers of 230, 200 and 190 kD and thus are named Tn230, Tn200 and Tn190, respectively. The difference in the sequence and experimentally determined molecular mass is mainly due to N-linked glycosylation for which Tn230 has 17 acceptor sites in contrast to the 11 of Tn200 and Tn190.

A model of the sequence organization of Tn230 is shown in Fig. 4, in which the

234

position of several recombinant and proteolytic fragments is also indicated.

Sequence Motifs of Laminin and Tenascin

The sequences of laminin and tenascin can be dissected into distinct domains. This becomes most obvious when the distribution of cysteines is observed. A dot matrix comparison clearly exhibits several internal homologies (Sasaki *et al.*, 1988). Although the entire sequences are unique extensive parts show clear similarities to other proteins. Starting from the amino-terminus in the following the major characteristics of each domain are discussed (cf. Figs. 3 and 4).

Amino-Terminal Domains

The N-terminal domain VI of mouse laminin is significantly conserved in all three chains and very similar to the other laminins sequenced so far. It consists of about 250 residues including 6 or 8 cysteines which can be aligned to identical positions (Sasaki *et al.*, 1988). The high similarity is further expressed by the tetrapeptide WWQS around position 80 to 100 and the unique pattern Y(Y/F)Yxhxdhxh(G/R)G terminating domain VI (h: hydrophobic residue; d: D, E or N). Domain VI corresponds to the globules at the tip of the short arms. Fragment C4 (similar to E4; Fig. 1c), comprising the outer region of the short arm build up by the B1-chain, is involved in the calcium-dependent stabilization of laminin and the calcium-dependent self-association (Bruch

---

[4]The amino termini of the B- and A-chains are located in the homologous terminal domains VI of the short arms (hatched). They are followed by tandemly repeating 8-Cys motifs (squares; domains V and III). In analogy to EGF the hypothetical structure of one repeat is shown in the upper right: parts which consist of an identical number of residues are shaded. Open circles mark loops of variable size. The specific arrangement of the repeats is indicated by numbers (identical numbers do not mean identity !). The inverse printed cysteines in the centre of the cross most probably form disulfide rings. Domains II and I forming the long arm build a triple stranded coiled-coil of α-helices. The differentially shaded boxes mark the alignment of heptad repeats. Coiled-coil factors F^n are calculated based on the values given in Table 1. The position of proline residues is marked by dots. The heptad repeats of the B1-chain are interrupted by domain α. Near the C-terminus the B-chains are disulphide linked (Cys printed inverse). The A-chain is continued by domain G consisting of five repeating units homologous to SHBG (cf. Fig. 5). Putative sites for N-glycosylation are marked by open triangles. For some sites evidence for glycosylation was obtained (filled triangles; Deutzmann *et al.*, 1988). The diameter of globular domains and the length of rods are calculated on the basis of the sequences (cf. scale bar). The position of different fragments is denoted by dashed lines. The N-termini of fragments E8, 25 K and E3 are shown on the right.

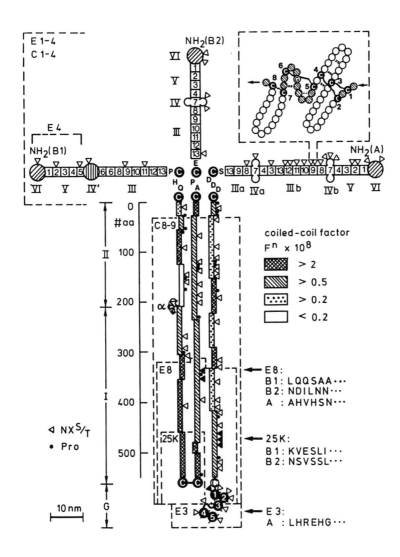

FIGURE 3. Structural model of mouse laminin based on sequence data[4].

236

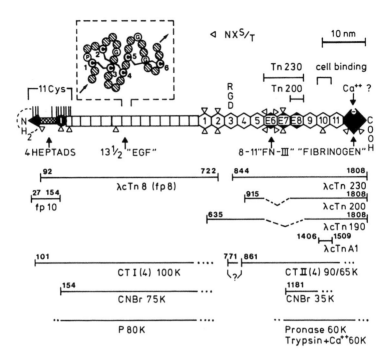

FIGURE 4. Structural model of chicken tenascin monomer.

The N-terminus of the tenascin monomers is located within the central globule and thus is drawn as a sector. It is followed by four heptad repeats (shaded box) and an intermediate domain I. The localization of cysteine residues including those of the half EGF-like repeat are marked by dashes and might be involved in the formation of interchain disulphide bonds. The thin proximal rods consist of 13 EGF-like repeats (squares) of which the hypothetical structure is shown as insert. The hexagons numbered 1 to 11 represent the fibronectin type III repeats which form the thick distal rods. The 'extra-domains' E6 to E8 and E8 are present only in the splicing variants Tn230 and Tn200, respectively. The 220 C-terminal residues are homologous to fibrinogen (see Fig. 6) and represent the distal globular domain (diamond). They contain a truncated EF-hand motif which could be responsible for calcium binding. Triangles mark the 17 putative acceptor sites for N-glycosylation. The length of the different domains is calculated on the basis of the sequences and are drawn to the scale.

Bars below the model indicate the position of several recombinant (λcTn, fp) and proteolytic fragments. Small numbers indicate the terminal residues. λcTnA1 contains a cell binding site and the epitope for anti-Tn68 (see Fig. 2g) whereas the RGD within the 3rd fibronectin type III repeat seems inactive. CT, CNBr and P denote fragments produced by chymotrypsin, cyanogen bromide and pepsin, respectively. Dots mark the range of uncertainty of their localization.

et al., 1989).

The N-terminal domain of mature tenascin consists of residue 34 to 118 and has no extensive homology to any known protein (Jones *et al.*, 1989; Spring *et al.*, 1989). It builds up the central globule of tenascin oligomers. This location has been confirmed by anti-Tn60 (Fig. 2e) which recognizes a fusion protein of residues 65–84 (Spring *et al.*, 1989). The three cysteines present in this region most probably are involved in interchain disulphide bonds. Residues 119 to 147 can be aligned as heptad repeats (see below) which are followed by a stretch of 30 residues which might form an independent domain.

Cysteine-Rich Domains

The sequences following domain VI of laminin are strikingly rich in glycine and cysteine. They can be aligned to tandem repeats with 8 cysteines in regular positions (Sasaki *et al.*, 1988). The regions along the first 6 cysteines have some homology to the epidermal growth factor (EGF). A most generalized consensus sequence of the laminin repeats compared to EGF reads:

```
             1    2        3        4      5                      6     7          8
Laminin:  x  C  x  C  x₆₋₁₄C  x₅₋₂₄C  x₁₋₂C x₂ g x₂ G x₂ C  x₂  C  x₉₋₂₃C  x

EGF    :  x₅C  x₇C  x₅       C  x₁₀   C  x    C x₂ G x₂ G x₂ C  x₆
```

If a disulphide linkage as in EGF is assumed the 7th and 8th cysteine could build up an additional loop. Drawn in analogy to EGF (Cooke *et al.*, 1987) in Fig. 3 the hypothetical structure of a laminin repeat is shown. This analogy might be justified as the crucial distance between the 5th and 6th cysteine including the glycine (underlined) is the same as in EGF (cf. Engel, 1989; 'g' denotes that this glycine is found in about half of the laminin repeats only).

The number of residues between the 8th to 2nd and the 5th to 7th cysteine is equal for all of the laminin Cys-rich repeats (shaded in Fig. 3). These residues probably build up the backbone of the short arms from which loops of variable size (open circles) are directed outward and thus determine the thickness of the rod.

The number of repeats is different for the B1- (13), B2- (11) and A-chain (17). An analysis of the repeats with respect to the size of the variable loops and the specific sequences suggests that they are arranged in a specific order. For example, the loop between the 2nd and 3rd cysteine of the repeats adjacent to domain VI consists of 6 residues, namely xGHASx and the consensus between the 5th and 6th cysteine reads xHNTxGxx. These characteristics are valid for all known laminin chains but are not found in any other repeat of the laminins. Based on such similarities (sometimes chosen not without arbitrariness) we have ordered the repeats as indicated by numbers (see Fig. 3).

In the B1-chains, the succession of 8-Cys repeats is interrupted by a unique domain IV' consisting of about 230 residues including 5 cysteines. On electron micrographs it probably corresponds to the inner globule of one of the 34 nm short arms. The adjacent Cys-rich repeats are also different from those of the B2- and A-chains: repeat "5" contains only 6 cysteines; the repeats "6" are similar to each other but dissimilar to those of the B2- and A-chains.

In the position equivalent to domain IV' the 8-Cys motifs of the B2- and A-chain, named 7, are unique in so far that the loopsize between the 3rd and 4th cysteine is increased to 180-200 residues lacking Cys. On electron micrographs therefore this repeat might be visualized as the inner globular domain of the other short arms. In the A-chain, repeat "7" and the neighbouring repeats "3/4" and "8/9" are duplicated. Therefore the 48 nm short arm with its additional globule near the centre of the cross (Fig. 1a) probably reflects the A-chain.

In contrast to the rapidly expanding group of proteins with EGF-like 6-Cys motifs, the 8-Cys laminin motif is less common. Interestingly it is found in another basement membrane component, namely the core protein of heparan sulfate proteoglycan (HSPG; Noonan *et al.*, 1988). When the 8-Cys motifs of the partial HSPG-sequence are compared with those of laminin their succession corresponds to "(7)-8-9-6-7-(8)" (brackets denote the limits of the available sequence).

In tenascin the intermediate domain I (Fig. 4) is followed by $13\frac{1}{2}$ tandem repeats 31 residues in length with 6 cysteines. Their consensus sequence in comparison to EGF reads:

$$
\begin{array}{llllllll}
 & 1 & 2 & 3 & 4 & 5 & & 6 \\
\text{Tenascin:} & x_2\ C\ P\ x_2 & C\ x_3\ G\ x\ C & x_4\ C\ x & \underline{C\ x_2\ G\ x_2\ G\ x_2\ C}\ x_2 \\
\text{EGF}\quad : & x_5\ C\ P\ x_6 & C\ x_3\ G\ x\ C & x_{10}C\ x & \underline{C\ x_2\ G\ x_2\ G\ x_2\ C}\ x_6
\end{array}
$$

The high degree of similarity to EGF especially within the loop between the 5th and 6th cysteine is obvious. Among the hundreds of 6-Cys EGF-like sequences the tenascin pattern is unique with respect to its shortness and its high internal homology (Spring *et al.*, 1989). When the different repeats are compared to each other the minimal and maximal values of identical residues are 15 and 28 out of 31, respectively. In Fig. 4 the hypothetical structure constructed in analogy to EGF (Cooke *et al.*, 1987) is shown as an insert.

These EGF-like repeats build up the proximal thin rods of tenascin. This is substantiated by the similar appearence of the fusion protein λcTn8 (Fig. 2h) which mainly consists of this part of the sequence (cf. Fig. 3).

The biological significance of the integration of growth factor related domains into large molecules presently is not understood. The specific order of the Cys-rich repeats in laminin might reflect their functional importance and argues against a simple structural role as spacer elements. Furthermore these domains are more strongly conserved during evolution than other domains of laminin (cf. Hunter *et al.*, 1989). For laminin it was shown that fragment P1 (Fig. 1d), which is mainly built up from 8-Cys repeats, has a mitogenic activity for different cells. This function is not correlated with the promotion of cell attachment (Panayotou *et al.*, 1989). Studies with synthetic peptides have identified a sequence YIGSR which is active in cell binding and is located in the backbone of the 8-Cys motif numbered "9" (Fig. 3) of the B1-chain (Graf *et al.*, 1987). In contrast to P1 the larger fragment E1-4, however, is inactive which might indicate a masking of this site by adjacent variable loops with different accessibilities for pepsin and elastase. For a similar reason the A-chain sequence RGD of repeat "4" in domain IIIb appears inactive in native laminin but is exposed in fragment P1 (Aumailley *et al.*, 1990). Also for intact tenascin it has been reported that it can promote the growth of mammary tumour cells (Chiquet-Ehrismann *et al.*, 1986). Whether this effect is due to the EGF-like repeats is not clear but recombinant tenascin fragments containing this region have been shown to be antiadhesive (Spring *et al.*, 1989). It has been proposed that growth factor related repeats might act as localized signals for growth and differentiation (Engel, 1989; Panayotou *et al.*, 1989).

Heptad Repeats (α-Helical Coiled-Coils)

The sequences of domain II and I which build up the long arm of laminin (see Fig. 3) consist of about 570 residues per chain and are delimited by cysteines. The N-terminal pairs of closely spaced cysteines probably form disulphide rings between the three chains in a similar arrangment as it has been proposed for the 'disulphide-knobs' at the borders of the fibrinogen coiled-coils (cf. Doolittle, 1984) and are located in the centre of the cross. The B-chains are disulphide-linked at the C-terminus (Paulsson *et al.*, 1985). Whereas the long arm sequence regions in the B2- and A-chain are not interrupted by cysteines domains II and I in the B1-chain are separated by a stretch of about 40 residues including 6 cysteines. These might form an independent domain designated α. The three chains have little sequence homology ($<$ 20 per cent identity) but a common pattern of hydrophobic residues. When written in groups of seven (abcdefg)$_n$, the a and d positions are preferentially occupied by hydrophobic amino acids. This sequence pattern is characteristic for proteins forming two- or three-stranded coiled-coils of α-helices (cf. Parry, 1982; Conway and Parry, 1990 and references therein).

In this structure residues a and d of one chain interact with those in position d and a of an adjacent chain thus forming a hydrophobic core. The shielding against the aqueous environment is the driving force for coiled-coil formation. This conformation can be stabilized by charged amino acids in opposite positions e and g. Polar residues in position b, c and f are exposed on the surface.

For an optimal adjustment of the entire sequences of domain II and I to the heptad scheme it is necessary to introduce several phase shifts. Frequently the edges coincide with proline residues which are known as α-helix breakers (dots in Fig. 3). Recently the disposition of amino acids within the heptads of several classes of double stranded α-helical coiled-coil proteins has been analyzed (Conway and Parry, 1990). Neglecting any specificity of the different proteins, from these values we have calculated the mean probability for the occurrence of an amino acid in a specific position (Table 1). Based on these data we determined a mean 'coiled-coil-factor' F^n for the several blocks of heptads (Fig. 3). F^n can be regarded as a measure for the probability that a given heptad array will form a coiled-coil (Parry, 1982).

The α-helical structure of the laminin long arm has been investigated in detail (Paulsson *et al.*, 1985; Bruch *et al.*, 1989). Circular dichroism spectra of the 25 k fragment whose sequences exhibit the highest value of F^n (Fig. 3) show extremely high negative amplitudes corresponding to about 100 per cent α-helical content. The combined data for fragments E8 and C8-9 (Fig. 1e–f; cf. Fig. 3) indicate that the α-helix is the prominent structural element of the entire long arm. The specific reassembly of denatured fragments E8 and C8-9 to complexes indistinguishable from the native fragments substantiate that the long arm is a triple-stranded coiled-coil (Hunter *et al.*, 1990). The specific nature of its formation also in the absence of the disulphide linkage suggests that it is the coiled-coil formation by which laminin is assembled in vivo. It was found that to a minor degree the B-chains alone can assemble to a two-stranded rope whereas the A-chain can participate in coiled-coil formation only in the

Percentage Occurence in the Heptad

Residue	residues/ 1000 res.	a	b	c	d	e	f	g
Ile	45	14.79	0.82	2.51	6.11	2.72	1.98	1.58
Phe	7	1.42	0.21	0.92	1.81	0.28	0.33	0.18
Val	37	7.25	1.69	2.52	8.28	1.08	2.53	2.21
Leu	126	29.54	2.79	3.42	38.22	4.33	3.14	6.88
Trp	1	0.06	0.02	0.03	0.60	–	0.07	0.02
Met	15	3.12	1.07	1.25	2.25	0.83	1.23	0.66
Ala	95	8.56	12.29	7.53	16.61	4.59	11.25	7.17
Glu	164	0.99	19.03	19.40	5.59	27.19	18.51	23.89
Asp	61	0.11	11.37	9.63	1.05	5.40	8.66	6.34
His	12	1.11	1.77	1.03	0.81	0.91	1.80	0.96
Arg	72	5.85	6.16	9.13	0.58	9.15	10.92	8.91
Lys	110	11.58	17.15	13.17	1.83	10.34	10.77	12.58
Gly	19	0.61	1.10	3.79	0.48	2.32	3.26	1.58
Tyr	14	3.21	0.34	0.13	5.12	0.11	0.10	0.09
Thr	46	1.07	2.71	4.49	2.99	7.73	7.17	5.67
Ser	48	1.42	5.49	5.12	3.73	4.89	6.10	6.85
Asn	46	7.33	6.20	6.74	0.64	4.51	4.90	2.36
Gln	77	0.67	9.68	8.86	3.04	13.40	6.34	11.87
Cys	4	1.30	0.03	0.32	0.25	0.25	0.76	0.19
Pro	1	0.01	0.09	0.02	0.04	–	0.17	0.01
apolar	326	64.74	18.98	18.18	63.88	13.83	20.53	18.70
acidic	225	1.10	30.40	29.03	6.64	32.59	27.17	30.23
basic	194	18.54	25.08	23.33	3.22	20.40	23.49	22.45

TABLE 1. Weighted fraction of residues occuring within the individual positions of the heptad repeats of double stranded coiled-coil proteins. This matrix is calculated from the values determined for the intermediate filament proteins (type I to V), myosin, tropomyosin, paramyosin, M6-protein and desmoplakin (Conway and Parry, 1990). The contribution from these six classes of proteins are scaled so as to be equal although M6-protein contributes only with 351 residues in contrast to the intermediate filament proteins where 13,184 residues were classified.

Polypeptid	M_r/L (nm^{-1})
Right-handed α-helix	750
two stranded α-helical coiled coil	1500
triple stranded α-helical coiled-coil	2250
single collagen helix	330
collagen triple helix	1000
fibronectin type I repeats (47 aa)	2300
fibronectin type II repeats (60 aa)	2700
fibronectin type III repeats (90 aa)	3500
6-Cys EGF-like repeats	1750
8-Cys Laminin-like repeats	2400

TABLE 2. Summary of mass-per-length ratios M_r/L for some polypeptide conformations and repeating sequence motifs.

As the Cys-rich and fibronectin-like motifs are modelled as compact spheres their M_r/L-ratios should be regarded as upper limits. Glycosylation effects have to be considered seperately.

presence of B-chains (I. Hunter and J. Engel, unpublished results). Interestingly this observation coincides with the difference in the average F^n-values which are 1.5×10^{-8} and 0.5×10^{-8} for the B- and A-chains, respectively (for comparison: the values of Table 2 result in $F^n \sim 5 \times 10^{-8}$ for nematode myosin).

In tenascin the amino-terminal globular domain is separated from domain I (Fig. 4) by a stretch of 29 amino acids which show a high tendency to form an α-helix according to secondary structure prediction. They can be aligned into four heptads with a coiled-coil factor of 3×10^{-8}. By assuming a similar mechanism as it has been worked out for laminin it can be hypothesised that for tenascin in a first step these short sequence regions could induce the assembly of a triple-stranded coiled-coil which then is stabilised by disulphide-rings at both ends. In tenascin oligomers this domain probably corresponds to the short rod connecting three short arms to the central globule (Spring *et al.*, 1989). This mechanism is suggestive to explain why in tenascin oligomers the number of chains mostly are multiples of three.

Fibronectin Type III Repeats of Tenascin

The EGF-like repeats of tenascin are followed by ~ 1000 amino acids (Tn230) without any cysteine. This part of the sequence can be divided into 11 repeated units about

90 residues in length with similarity to the fibronectin type III motif. The perfectly conserved residues within this alignment are only a tryptophan around the 25th, a leucine around the 65th and a threonine at the last position (Spring *et al.*, 1989). The tenascin variants resulting from alternative splicing differ in the number of these motifs. In Tn200 the 6th and 7th and in Tn190 the 6th to 8th repeat are deleted (black hexagons in Fig. 4). The splicing out occurs exactly at the borders of the repeats.

The sequence identity between the several repeats varies between 12 to 46 per cent. The 'extra-domains' E6 to E8 are most similar to each other (41 to 46 per cent identity) but most dissimilar to the others. When compared to the type III domains of chicken fibronectin the sequence identity varies between 13 to 40 per cent. No obvious similarity of the repeat arrangement could be detected between these two proteins.

As substantiated by the specific recognition of monoclonal antibody epitopes and the equivalent appearence of fusion proteins (Fig. 2f and i) the thick distal portion of tenascin is built up by these fibronectin type III repeats. A strong cell binding site is located within the 10th (or at the edge of the 11th) repeat whereas the RGD-sequence present within the third repeat appears to be inactive (Spring *et al.*, 1989; Chiquet-Ehrismann, 1991).

In contrast to the EGF-motif and also the heptad repeat pattern it is much more difficult to define the fibronectin type III motif. As a general characterisation it should appear as a distinct (not necessarily repeating) unit about 80 to 100 residues in length including certain conserved aromatic and nearby hydrophobic residues. These conditions are fulfilled by a rapidly growing number of proteins including extra-, intracellular and also membrane proteins (cf. Norton *et al.*, 1990; Patthy, 1990). When the tenascin repeats are compared to related repeats of several other proteins, it is found that the mean similarity is highest for those of fibronectin. Whether this structural relation is the basis for the competition of these two extracellular matrix proteins in several of their biological functions remains to be determined (cf. Chiquets-Ehrismann *et al.*, 1988; Chiquet-Ehrismann, 1991).

As far as structural information is available, fibronectin type III repeats are likely to be arranged in a linear array thus forming rod-shaped structures of considerable flexibility. On electron micrographs the flexibility is emphasized by a remarkable bending of the rods (see e.g. Engel *et al.*, 1981 for fibronectin; Nave *et al.*, 1989 for titin which also contains such repeats: Labeit *et al.*, 1990).

Homology of Sex Hormone Binding Globulin to C-Terminus of Laminin A-Chain

Domain G of the A-chain has no counterpart within the B-chains and most probably it builds up the distal globular structure of the long arm (cf. Fig. 3). The sequence can be divided into five 150 to 180 residue repeats Lam-G1 to Lam-G5 (Fig. 5) which might represent the five subdomains detected by negative stain electron microscopy. Whereas the sequence identity is only 15 to 31 per cent several glycines and prolines are found at regular intervals and also the location of hydrophobic and charged residues is highly conserved. Especially a pair of cysteines is placed in corresponding positions (Deutzmann *et al.*, 1988; Sasaki *et al.*, 1988). For G5 the mode of disulphide linkage has been been elucidated by biochemical analysis (Fig. 5). The cluster of basic amino acids around the cysteines in G5 is not found in the other repeats and probably represents a major heparin binding site of laminin which might be involved in the binding of HSPG (Timpl *et al.*, 1990).

A homology search for domain G reveals its highest similarity to the sex hormone binding globulin (SHBG) of human. Interestingly for the different subdomains of laminin different parts of SHBG were aligned by the computer. When analysed in more detail it becomes evident that SHBG itself seems to consist of two similar domains SHBG-1 and SHBG-2 as schematically drawn in Fig. 5. The most siginificant features of this alignment are the conserved positions of the C-terminal cysteines and several blocks around glycine and proline residues. The disulphide linkage of SHBG is not known but from its similarity to the C-terminal domain of protein S it was suggested that Cys_{164}/Cys_{188} and Cys_{333}/Cys_{361} form disulphides (Gershagen *et al.*, 1987). By analogy to laminin the same disulphide bonds could be expected although the protein S homology does not hold for laminin. Also the steroid binding site of SHBG is unknown but might be located between residues 240 to 290. The alternating leucine pattern of this region is not found in laminin. When the different subdomains of laminin are compared to those of SHBG, the highest similarities are found to alternate between SHBG-1 and SHBG-2 (arrows in table of Fig. 5) which becomes more obvious when several conserved blocks are compared (e.g. "SASA" in SHBG-1 and Lam-G4; "HSCP" in SHBG-2 and Lam-G5).

At present no data are available which would indicate a steroid binding activity of domain G in laminin. The long arm terminal domain was found to play a crucial role in developmental regulation. Antibodies specific for fragment E3 corresponding to regions G4/G5 (cf. Fig. 3) can inhibit the polarization of developing cells. In organ

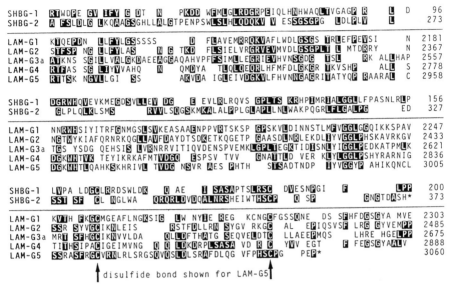

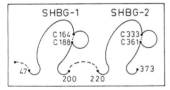

FIGURE 5. Sequence alignment of domain G of mouse laminin A-chain and human sex hormone binding globulin (SHBG).

The major part of the SHBG sequence can be divided in two related domains SHBG-1 and SHBG-2 (see cartoon). The co-alignment of the internal repeating units G1 to G5 of laminin exhibits the similar position of cysteines (arrows) and several conserved blocks especially around glycine and proline residues. Amino acids which are identical in at least one domain G_i with either SHBG-1 or SHBG-2 are printed inverse. The table shows the sequence identity within this alignment. Arrows denote that the succession of G_i appears similar to the SHBG domains.

cultures of embryonic kidney the A-chain is expressed just at the onset of conversion of mesenchyme to epithelium whereas the B-chains are already synthesized by nonpolar cells (cf. Ekblom, 1989 and references therein).

Fibrinogen Homology to C-Terminus of Tenascin

The distal globule of the tenascin arms is formed by the C-terminal 220 amino acids which show homology to another rapidly growing family of proteins with related domains, the fibrinogen-like sequences. The distal globular domains of the β- and γ-chains of fibrinogen were for a long time the only members of this family (Doolittle, 1984). But since the β- and the γ-chain are only about 35 per cent identical, while the corresponding chains of lamprey and man are 50 per cent identical, it was anticipated that a common ancestor should exist in invertebrates. Indeed, two such sequences were found in a sea cucumber and one was reported to exist in drosophila (Xu and Doolittle, 1990).

Fig. 6 shows an alignment of some fibrinogen-like sequences with the β- and γ-chain of lamprey (FBEL and FGAL) and the γ-chain of human fibrinogen (FGAH). The highest degree of identity (52 %) with chicken tenascin (TNCH) can be seen in the human sequence (URFH) found as an anti-sense transcript of the steroid 21-hydroxylase/complement component C4 gene locus (Morel et al., 1989). No gaps have to be introduced in relation to tenascin and its similarity extends further into the fibronectin type III repeats towards the N-terminus. The similarity of URFH with chicken tenascin is slightly higher than with the partial sequence of human tenascin (not shown). Therefore, these gene products might be considered as tenascin-1 and tenascin-2, respectively. A protein deduced from a transcript specific for killer T-cells (FCYM) shows homology to the fibrinogens over the entire length but lacks the functional hallmarks (Koyama et al., 1987). The identity with tenascin (39 %) is comparable to those with the fibrinogens but is restricted to the globular part of FCYM which was shown to be encoded by a separate exon. The sea cucumber sequence (FGSC; Xu and Doolittle, 1990), however, shows a significantly higher degree of homology to tenascin (46 %) than to the fibrinogens. As this sequence is not part of a multidomain protein it could represent the descendant of the original 'fibrinogen-globule'. Although Xu and Doolittle favour the idea that this protein would form dimers, based on the homology to the fibrinogens, the pictures of single globules at the tips of the tenascin arms might indicate that fibrinogen-globules could exist as monomers.

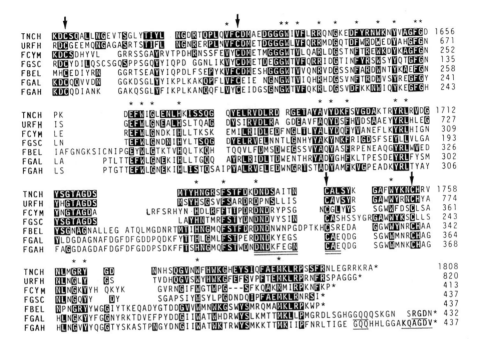

FIGURE 6. Comparison of the C-terminal domain of tenascin with proteins homologous to fibrinogen.

The C-terminal sequence of chicken tenascin (TNCH) is compared to a human unidentified reading frame (URFH; Morel *et al.*, 1989), the cytotoxic T-lymphocyte-specific transcript pT49 from mouse (FCYM), a fibrinogen-like protein from sea cucumber (FGSC) and the globular part of the fibrinogen β-and γ-chain of lamprey (FBEL, FGAL) and human (FGAH; for references see Xu and Doolittle, 1990). Residues identical to tenascin are printed inverse. Asterisks mark residues identical to all proteins. Common cysteines are highlighted by arrows. The EF-hand of the calcium binding site in FGAH is double underlined. The transglutamination site GQQ of the γ-chains, the RGD in FGAL, and the hexapeptide of FGAH competing for RGD binding are underlined to emphasize their absence in tenascin.

The four cysteines which are disulphide-linked in fibrinogen are perfectly conserved in all these sequences. Therefore it can be hypothesized that in tenascin intrachain disulphide loops might be formed between Cys_{1591}/Cys_{1621} and Cys_{1743}/Cys_{1756}. The EF-hand motif shown to be an active calcium binding site in human γ-fibrinogen (Dang et al., 1985) is interrupted in β-fibrinogen but reasonably well conserved in tenascin. Whether it represents the claimed calcium binding site of tenascin (Jones et al., 1989) is presently unclear. Although in most cases several EF-hand motifs are repeating in the sequence and pairwise packing is observed in three-dimensional structures, single isolated EF-hands are also found in α-lactalbumin and the extracellular glycoprotein BM-40/SPARC/osteonectin (Engel et al., 1987 and references therein). A crucial deviation from the consensus of a calcium-binding EF-hand, however, is found in the (-z)-position of the loop (Szebenyi and Moffat, 1986) where tenascin contains leucine instead of glutamic or aspartic acid.

The C-terminus of γ-fibrinogen contains a specific sequence for transglutamination (Doolittle, 1984). The very last six amino acids of the human γ-fibrinogen, KQAGDV, were shown to compete with RGD-peptides for binding to integrins (Hautanen et al., 1989); interestingly, the lamprey γ-fibrinogen contains an RGD-sequence at this position. Both sites, however, are not present in tenascin or the other members of this family.

"Molecular Modelling" by Motif Analysis and Electron Microscopy

Due to the extended and branched shape of laminin and tenascin their atomic structure will not be available in the near future. The correlation of defined sequence elements with the shape and dimensions determined by electron microscopy, however, enables a rather precise mapping of specific epitopes. The models of laminin and tenascin presented in Figs. 3 and 4 have been constructed on the basis of the sequences. The sequence motifs discussed above have been assumed as individual structural elements arranged in a linear array like beads on a string. For a compact spherical molecule the molar mass M_r is related to the diameter d by

$$d = (6M_r\, v_2/\pi N_A)^{1/3}$$

where N_A is Avogadro's number and v_2 the partial specific volume.

For the Cys-rich repeats of laminin and tenascin an average translation per domain of 2.4 nm and 2.0 nm, respectively, result which agrees well with the dimensions of the rod-like regions determined by electron microscopy and the predicted dimensions assuming an EGF-like conformation (Cooke *et al.*, 1987; Engel, 1989). Both for the fibrinogen-like domain of tenascin as well as domains VI and IV of laminin the diameters are calculated to 3.9 nm. By negative stain electron microscopy the short arm globules were measured to 5 ± 1 nm (Engel *et al.*, 1981). For the fibronectin type III motif a diameter of 2.9 nm is predicted which fits well with the length of the thick distal rod of tenascin. Especially the epitopes of antibodies specific for the different splicing variants are located in the expected distance from the distal globular domain (Spring *et al.*, 1989). For the fibronectin type III related repeats of titin the diameter is assumed to 4.3 nm (cf. Labeit *et al.*, 1990; Nave *et al.*, 1989). Whether this difference reflects an ellipsoid shape of these domains and/or results from different degrees of stretch exerted on this probably rather elastic molecule remains unclear. The five repeats of domain G are modelled by globules 3.6 nm in diameter which must be compared to the 4.0 ± 0.6 nm determined for the subdomains of the distal globule of the laminin long arm (Bruch *et al.*, 1989).

The heptad repeat pattern is modelled by cylinders assuming an α-helical structure with a mean helix rise of 0.15 nm/residue. For tenascin the length of the short helix between the central and intermediate domains should measure 4 nm and domains II/I of laminin should extend over 85 nm. For human hexabrachion/tenascin a distance between the central knob and the connection of three arms has been estimated to 2–5 nm (Erickson and Inglesias, 1984). The laminin long arm measures 77 nm (Engel *et al.*, 1981). The difference for laminin might be due to helical interruptions (phase shifts) frequently introduced by proline residues which could build short loops resulting in a more compact structure.

According to this sequence modelling the length of a tenascin monomer is 64.6 nm (Tn190), 67.5 nm (Tn200) and 73.3 nm (Tn230) (Spring *et al.*, 1989) which corresponds well with the 66 nm to 75 nm determined by several groups (Chiquet-Ehrismann *et al.*, 1988; Erickson and Bourdon, 1989 and references therein). The sum over the short arm sequence elements of laminin results in 39.0 nm (B1), 30.3 nm (B2) and 44.7 nm (A) which has to be compared with 34 ± 4 nm (mean of two short arms) and 48 ± 4 nm (Bruch *et al.*, 1989), respectively.

In summary the model data for the rod-like elements are rather consistent with those determined by electron microscopy and similar good correlations have been found for several other molecules including fibronectin, nidogen and vinculin (cf. Engel and Furthmayr, 1987; Beck, 1989). In an initial stage of the analysis of an extended molecule it might be helpful to correlate specific sequence elements to locations on electron micrographs. For this purpose in Table 2 the mass-per-length ratios of some polypeptide conformations are summarized.

The analysis of the structure and function of laminin and tenascin demonstrates that they are constructed by the repetition of small domains which together form a multifunctional protein. The difference of the three tenascin variants shows that sequence motifs are correlated with biological processes. The similarity of some recombinant fragments with the corresponding parts of native tenascin (Fig. 2h–i) suggests an independent folding of the different domains (Spring *et al.*, 1989). The perfect renaturation of laminin fragments which also occours after reduction and alkylation demonstrates the highly specific nature of coiled-coil formation and the rearrangment of the long arm globular domains (Hunter *et al.*, 1990). These observations suggest that sequence homology is translated into structural homology. Sequence regions belonging to the same type of motifs probably form domains of very similar three-dimensional structure. Therefore a deeper insight into the structure of such multidomain proteins can be expected from a systematic analysis of small characteristic domains by X-ray or NMR techniques. Presently only a few pieces for this puzzle are available but research in this direction has already started (cf. Holland and Blake, 1989).

Acknowledgements: Original work reviewed in this paper was in part supported by The Swiss National Science Foundation. K.B. was the recipient of a research fellowship from the Deutsche Forschungsgemeinschaft.

References

Aumailley, M., Gerl, M., Sonnenberg, A., Deutzmann, R., and Timpl, R. (1990). Identification of the Arg-Gly-Asp sequence in laminin A chain as a latent cell-binding site being exposed in fragment P1. Identification of the Arg-Gly-Asp sequence in laminin A chain as a latent cell-binding site being exposed in fragment P1. *FEBS Lett.*, 262:82–86.

Beck, K. (1989). Structural model of vinculin: correlation of amino acid sequence with electron-microscopical shape. *FEBS Lett.*, 249:1–4.

Beck, K., Hunter, I., and Engel, J. (1990). Structure and function of laminin: anatomy of a multidomain glycoprotein. *FASEB J.*, 4:148–160.

Bruch, M., Landwehr, R., and Engel, J. (1989). Dissection of laminin by cathepsin G into its long arm and short arm structures and localization of regions involved in calcium dependent stabilization and self-association. *Eur. J. Biochem.*, 185:271–279.

Chiquet, M. (1989). Tenascin/J1/cytotactin: the potential function of hexabrachion proteins in neural development. *Dev. Neurosci.*, 11:266–275.

Chiquet, M., Schenk, S., Beck, K., Nowotny, N., and Chiquet-Ehrismann, R. (1991). Protein domains of tenascin: the C-terminal 60k fragment binds to heparin and preferentially arise from the large isoform. *J. Biol. Chem.* Submitted.

Chiquet-Ehrismann, R. (1991). What distinguishes tenascin from fibronectin? *FASEB J.*, 4(8).

Chiquet-Ehrismann, R., Kalla, P., Pearson, C. A., Beck, K., and Chiquet, M. (1988). Tenascin interferes with fibronectin action. *Cell*, 53:383–390.

Chiquet-Ehrismann, R., Mackie, E. J., Pearson, C. A., and Sakakura, T. (1986). Tenascin: an extracellular matrix protein involved in tissue interactions during fetal development and oncogenesis. *Cell*, 47:131–139.

Conway, J. F. and Parry, D. A. D. (1990). Structural features in the heptad substructure and longer range repeats of two stranded α-fibrous proteins. *Int. J. Biol. Macromol.*, 12:328–334.

Cooke, R. M., Wilkinson, A. J., Baron, M., Pastore, A., Tappin, M. J., Campbell, I. D., Gregory, H., and Sheard, B. (1987). The solution structure of human epidermal growth factor. *Nature*, 327:339–341.

Dang, C. V., Ebert, R. F., and Bell, W. R. (1985). Localization of a fibrinogen calcium binding site between γ-subunit positions 311 and 336 by terbium fluorescence. *J. Biol. Chem.*, 260:9713–9719.

Deutzmann, R., Huber, H., Schmetz, K. A., Oberbäumer, I., and Hartl, L. (1988). Structural study of long arm fragments of laminin. evidence for repetitive C-terminal sequences in the A-chain, not present in the B-chains. *Eur. J. Biochem.*, 177:35–45.

Doolittle, R. F. (1984). Fibrinogen and fibrin. *Annu. Rev. Biochem.*, 53:195–229.

Ekblom, P. (1989). Developmentally regulated conversion of mesenchyme to epithelium. *FASEB J.*, 3:2141–2150.

Engel, J. (1989). EGF-like domains in extracellular matrix proteins: localized signals for growth and differentiation? *FEBS Lett.*, 251:1–7.

Engel, J. and Furthmayr, H. (1987). Electron microscopy and other physical methods for the characterization of extracellular matrix components: laminin, fibronectin, collagen IV, collagen VI, and proteoglycans. *Methods Enzymol.*, 145:3–78.

Engel, J., Odermatt, E., Engel, A., Madri, J. A., Furthmayr, H., Rohde, H., and Timpl, R. (1981). Shapes, domain organization and flexibility of laminin and fibronectin, two multifunctional proteins of the extracellular matrix. *J. Mol. Biol.*, 120:97–120.

Engel, J., Taylor, W., Paulsson, M., Sage, H., and Hogan, B. (1987). Calcium binding domains and calcium induced conformational transition of SPARC/BM-40/osteonectin, an extracellular glycoprotein expressed in mineralized and non-mineralized tissues. *Biochemistry*, 26:6958–6965.

Erickson, H. P. and Bourdon, M. A. (1989). Tenascin: an extracellular matrix protein prominent in specialized embryonic tissues and tumors. *Annu. Rev. Cell Biol.*, 5:71–92.

Erickson, H. P. and Inglesias, J. L. (1984). A six-armed oligomer isolated from cell surface fibronectin preparations. *Nature*, 311:267–269.

Erickson, H. P. and Lightner, V. A. (1988). Hexabrachion protein (tenascin, cytotactin, brachionectin) in connective tissues, embryonic brain and tumors. In Miller, K. R., editor, *Advances in Cell Biology*, pages 55–90. London, JAI.

Friedlander, D. R., Hoffman, S., and Edelman, D. M. (1988). Functional mapping of cytotactin: proteolytic fragments active in cell substrate adhesion. *J. Cell Biol.*, 107:2329–2340.

Gershagen, S., Fernlund, P., and Lundwall, Å. (1987). A cDNA coding for human sex hormone binding globulin. homology to vitamin K-dependent protein S. *FEBS Lett.*, 220:129–135.

Graf, J., Iwamoto, Y., Sasaki, M., Martin, G. R., Kleinman, H. K., Robey, F. A., and Yamada, Y. (1987). Identification of an amino acid sequence in laminin mediating cell attachment, chemotaxis and receptor binding. *Cell*, 48:989–996.

Hautanen, A., Gailit, J., Mann, D. M., and Ruoslahti, E. (1989). Effects of modifications of the RGD sequence and its context on recognition by the fibronectin receptor. *J. Cell Biol.*, 264:1437–1442.

Holland, S. K. and Blacke, C. C. F. (1989). Multi-domain proteins: towards complete structures. In Aebi, U. and Engel, J., editors, *Cytoskeletal and Extracellular Proteins*, pages 137–139. Heidelberg: Springer-Verlag.

Hunter, D. D., Shah, V., Merlie, J. P., and Sanes, J. R. (1989). A laminin-like adhesive protein concentrated in the synaptic cleft of the neuromuscular junction. *Nature*, 338:229–234.

Hunter, I., Schulthess, T., Bruch, M., Beck, K., and Engel, J. (1990). Evidence for a specific mechanism of laminin assembly. *Eur. J. Biochem.*, 188:205–211.

Jones, F. S., Hoffman, S., Cunningham, B. A., and Edelman, G. M. (1989). A detailed structural model of cytotactin: protein homologies, alternative splicing, and binding regions. *Proc. Natl. Acad. Sci. USA*, 86:1905–1909.

Koyama, T., Hall, L. R., Haser, W. G., Tonegawa, S., and Saito, H. (1987). Structure of a cytotactic T-lymphocyte-specific gene shows a strong homology to fibrinogen β and γ chains. *Proc. Natl. Acad. Sci. USA*, 84:1609–1613.

Labeit, S., Barlow, D. P., Gautel, M., Gibson, T., Holt, J., Hsieh, C.-L., Francke, U., Leonard, K., Wardale, J., Whiting, A., and Trinick, J. (1990). A regular pattern of two types of 100-residue motif in the sequence of titin. *Nature*, 345:273–276.

Masuda-Nakagawa, L., Beck, K., and Chiquet, M. (1989). Identification of molecules in leech extracellular matrix that promote neurite outgrowth. *Proc. R. Soc. Lond. B*, 235:247–257.

Morel, Y., Bristow, J., Gitelman, S. E., and Miller, W. L. (1989). Transcript encoded on the opposite strand of the human steroid 21-hydroxylase/complement component C4 gene locus. *Proc. Natl. Acad. Sci. USA*, 86:6582–6586.

Nave, R., Fürst, D. O., and Weber, K. (1989). Visualization of the polarity of isolated titin molecules: a single globular head on a long thin rod as the M-band anchoring domain? *J. Cell Biol.*, 109:2177–2187.

Noonan, D. M., Horigan, E. A., Ledbetter, S. R., Vogeli, G., Sasaki, M., Yamada, Y., and Hassell, J. R. (1988). Identification of cDNA clones encoding different domains of the basement membrane heparan sulfate proteoglycan. *J. Biol. Chem.*, pages 16379–16387.

Norton, P. A., Hynes, R. O., and Rees, D. J. G. (1990). Sevenless: seven found? *Cell*, 61:15–16.

Panayotou, G., End, P., Aumailley, M., Timpl, R., and Engel, J. (1989). Domains of laminin with growth-factor activity. *Cell*, 56:93–101.

Parry, D. A. D. (1982). Coiled-coils in α-helix containing proteins: analysis of the residue types within the heptad repeat and the use of these data in the prediction of coiled-coils in other proteins. *Biosci. Rep.*, 2:1017–1024.

Patthy, L. (1990). Homology of a domain of the growth hormone/prolactin receptor family with type III modules of fibronectin. *Cell*, 61:13–14.

Paulsson, M., Deutzmann, R., Timpl, R., Dalzoppo, D., Odermatt, E., and Engel, J. (1985). Evidence for coiled-coil α-helical regions in the long arm of laminin. *EMBO J.*, 4:309–316.

Sasaki, M., Kleinman, H. K., Huber, H., Deutzmann, R., and Yamada, Y. (1988). Laminin, a multidomain protein. the A chain has a unique globular domain and homology with the basement membrane proteoglycan and the laminin B chains. *J. Biol. Chem.*, 263.

Spring, J., Beck, K., and Chiquet-Ehrismann, R. (1989). Two contrary functions of tenascin: dissection of the active sites by recombinant tenascin fragments. *Cell*, 59:325–334.

Szebenyi, D. M. E. and Moffat, K. (1986). The refined structure of vitamin D-dependent calcium-binding protein from bovine intestine. Molecular details, ion binding, and implications for the structure of other calcium-binding proteins. *J. Biol. Chem.*, 261:8761–8777.

Timpl, R. (1989). Structure and biological activity of basement membrane proteins. *Eur. J. Biochem.*, 180:487–502.

Timpl, R., Aumailley, M., Gerl, M., Mann, K., Nurcombe, V., Edgar, D., and Deutzmann, R. (1990). Structure and function of the laminin-nidogen complex. *Ann. N. Y. Acad. Sci.*, 580:311–323.

Xu, X. and Doolittle, R. F. (1990). Presence of a vertebrate fibrinogen-like sequence in an echinoderm. *Proc. Natl. Acad. Sci. USA*, 87:2097–2101.

A Sequence Motif in the Transmembrane Region of Tyrosine Kinase Growth Factor Receptors

Michael J. E. Sternberg[1]and William J. Gullick[2]

Biomolecular Modelling Laboratory[1]
Imperial Cancer Research Fund
P.O. Box 123
Lincoln's Inn Fields
London WC2A 3PX
U.K.

m_sternberg@uk.ac.icrf

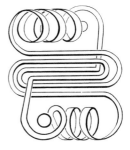

Introduction

The tyrosine kinase (TK) family of growth factor receptors (GFR) consists of one or more polypeptide chains organised into three regions; an N-terminal extracellular domain followed by a transmembrane section that leads to the intracellular domain e.g. (Hanks *et al.*, 1988; Gullick, 1988). Ligand binding to the extracellular domain conveys the signal to the intracellular domain to induce its TK activity. This increase in TK activity then conveys the mitogenic stimulus to the nucleus promoting cell growth and division. The mechanism by which extracellular ligand binding stimulates intracellular TK activity is uncertain but there is considerable indirect evidence (e.g. Yarden and Schlessinger, 1987) that binding alters the conformation of the extracellular domain and promotes dimerisation of the receptors.

There is close association between GFRs and cancer. In certain types of cancer, such as some breast and other cancers, (Slamon *et al.*, 1987; Sainsbury *et al.*, 1987), there is an overexpression of some TK-GFRs such as the EGF-receptor and c-erbB-2 protein that are considered to promote unregulated growth. In addition, certain oncogenes (Hanks *et al.*, 1988) are altered GFRs with increased, ligand-independent TK activity that can promote tumour formation (Here these oncogenes are also referred to as GFRs).

Recently insight into the possible mechanism of TK activation was obtained from

---

[2]Molecular Oncology Laboratory, Imperial Cancer Research Fund, Oncology Group, Hammersmith Hospital, Du Cane Road, London W12 OHS, U.K.

an observation on one member of the TK-GFR family, the rat neu oncogene. Rat neu (Bargmann *et al.*, 1986) is the homologue of the human c-erbB-2 and both these proteins are related in sequence to both the extracellular domain and the TK domain of the EGF receptor. A single carcinogen-induced, point mutation (Bargmann and Weinberg, 1988) at a specific position in its transmembrane region (Val 664 → Glu) leads to conversion of c-neu (the cellular proto-oncogene) into onc-neu (the oncogene) that possesses a considerably increased TK activity and is transforming. We report a proposal (Gullick, 1988; Sternberg and Gullick, 1989, 1990) that the Glu mutation in onc-neu promotes the dimerisation of transmembrane α-helices leading to GFR association and TK activity. It has subsequently been demonstrated (Weiner *et al.*, 1989) that onc-neu, but not c-neu, predominantly associates into dimers. Examination of the sequences of the transmembrane region of all 20 known GFRs reveals a sequence motif related to the specific packing in neu suggests that specific dimerisation may be a more general phenomenon.

Neu Dimerisation

Fig. 1 shows our specific model for dimerisation (for a space filling diagram of the helix packing see Fig. 1 of Sternberg and Gullick, 1989). The stereochemistry of this proposal was proposed from studies using both space filling molecular models and docking on computer graphics.

The conformation of the transmembrane region was taken as an α-helix. The hydrophobic environment of the membrane (Engleman and Steitz, 1981) raises the pK_a of Glu in onc-neu so the majority of these side chains will be protonated and capable of forming hydrogen bonds at physiological pH. It is proposed that the carboxyl group of Glu 664 in one helix (A) forms a hydrogen bond with the carbonyl oxygen of Ala 661 in the other helix (B) (see Fig. 1a). The carbonyl oxygen of Ala 661 still forms the main-chain / main-chain hydrogen bonding for the α-helix. Such bifurcated hydrogen bonding (Baker and Hubbard, 1984) is common within the α-helices with the hydroxyl side chains of Ser and Thr and can occur for Glu. A second and symmetric hydrogen bond is formed between Glu 664 in B with the oxygen of Ala 661 in A. The helical axes lie at about −50° which is a favourable orientation for α-helix packing (Chothia *et al.*, 1981). This angle would provide some separation for dimerisation of the extracellular and of the intracellular domains thus avoiding steric hindrance.

In the c-neu protein (Fig. 1b) the same helix/helix arrangement would pack Val 664

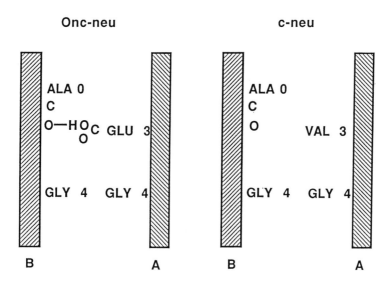

FIGURE 1. A schematic diagram of the packing of the transmembrane α-helices in the onc-neu and c-neu dimers.

α-helices A and B are denoted by diagonally shaded rectangles. The numbering of the residues corresponds to that in the motif. Fig. 1a (left) shows onc-neu with the formation of a hydrogen bond between the COOH of Glu in helix B with the main chain carbonyl oxygen in helix A. The close packing of Gly in both helices is shown. Fig. 1b (right) illustrates that in c-neu the side chain of Ala in helix B will pack in the space between Ala and Gly in helix A. From Sternberg and Gullick (1990).

(denoted position P3) in helix A against Gly 665 (P4) in helix B. This model for c-neu requires a sequence specific pattern with an aliphatic side chain at position 3 (denoted P3) (Val 664 in neu) packing against small side chains at P0 and P4 (Ala 661, Gly 665). This model explains TK activation by onc-neu in terms of the increased stability of the dimer compared to that in c-neu.

There is an alternative model for neu dimerisation in which the two transmembrane α-helices do not pack but the side chain of Glu 664 is extended in helices A and B and a Glu–Glu hydrogen bond is formed thus stabilising the onc-neu dimer compared to c-neu.

Sequence Motif

The transmembrane sequences of all 20 known TK-GFRs were examined to establish the generality of this sequence pattern (Tab. 1). Inspection of the sequences showed that a generalised version of the P0, P3 and P4 residues in neu were found at least once in 18 out of the 20 transmembrane sequences. P0 requires a small side chain (Gly, Ala, Ser, Thr or Pro); P3 an aliphatic side chain (Ala, Val, Leu or Ile) and P4 only the smallest side chains (Gly or Ala). The strong restriction on residues at P4 suggests that this side chain packs directly against that of P3 with the constraint on P0 being that the side chain is not too large to introduce steric hindrance. The presence of the motif in 18 sequences suggests that helix association is a general mechanism for TK activation in these receptors.

In certain GFRs, the motif occurs more than once. The motif boxed in table 1 was chosen to maintain a similar location within the transmembrane region to that observed in neu. However in 3GFRs (v-erbB, Trk and Ltk) this still left an arbitrary assignment on the location of the motif. The absence of the pattern in v-fms might reflect a different mechanism of TK activation due to carboxyl terminal mutation compared to its cellular counterpart, the CSF1-R (c-fms) (Coussens *et al.*, 1986). It is not clear why this pattern is not observed in the drosophila EGF-receptor homologue (Livneh *et al.*, 1985).

The significance of finding this pattern in 18 out of the 20 sequences was evaluated following the approach of Rothbard and Taylor (1988). From the frequency of occurrence of the residue types in the hydrophobic region, the probability for the residues occurring at the P0, P3 and P4 are 0.311, 0.680 and 0.169. Thus the probability (p) of observing the pattern at one location is 0.0358. There are 21 locations where the pattern could occur within the typical 25 residue hydrophobic segment, so the pattern

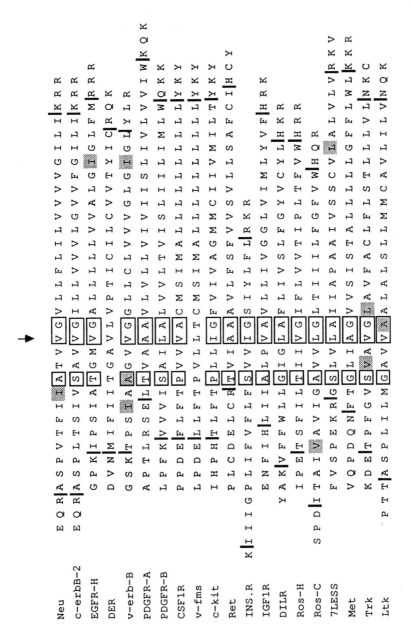

TABLE 1. The sequence pattern in the transmembrane domain of tyrosine kinase growth factor receptors. One set of residues forming the pattern of positions P0, P3 and P4 are boxed. Further occurrences in a sequence are denoted by background shading at P3. The arrow points to Val 664 (V) in neu to denote the position of the mutations that lead to TK activation. The N- and C-termini of the hydrophobic section of the transmembrane region are denoted by bold vertical lines. For details of the proteins included see Sternberg and Gullick (1990) on which the table is based.

is expected not to occur in one sequence with a probability of $(1 - p)^{21}$. Thus the probability that it occurs *at least once* is $1 - (1 - p)^{21} = 0.535$. The likelihood of observing this pattern 18 or more times in 20 sequences is evaluated from a one-tailed binomial distribution as 0.0006.

This type of probability analysis does not consider that the motif may be a general feature in the sequence of transmembrane regions. To investigate this one must examine the sequences of transmembrane regions that are not from TK-GFRs. There are several classes of transmembrane regions and GFRs belong to the class of proteins with a cleavable signal peptide and one internal membrane segment (von Heijne and Gavel, 1988). Accordingly, a reference set of 33 non homologous transmembrane sequences from non-viral eukaryotic proteins belonging to this class was obtained from the OWL database (Akrigg *et al.*, 1988). In the 33 reference protein chains, the motif does not occur in 13 transmembrane regions whereas in the 20 GFRs the motif does not occur in 2 transmembrane regions. A χ^2-test shows that these distributions are significantly different at the 2.5% level. Thus although the motif is not restricted to the GFRs, it does occur far more often in these proteins than in the transmembrane regions of other proteins.

Conclusion

The presence of the proposed helix packing motif at P0, P3 and P4 in the transmembrane region of most members of the TK-GFR family is unlikely to be the result of chance. This suggests that a dimerisation involving the close packing of α-helices similar to that proposed for neu may occur in many other TK-GFRs as a result of ligand binding or increased receptor expression. However, at this stage one cannot rule out the alternative model involving dimer stabilisation between α-helices that do not pack. Nevertheless, this analysis suggests that the transmembrane region of GFRs may well have a central role in dimerisation and should not be regarded simply as a passive spacer.

If this model of helix association is correct, then peptides with the sequence of the transmembrane region of one member of the family might be able to compete with other receptor molecules of the same family during dimerisation. If the affinity of the peptide for the receptor is sufficiently high, then the peptide might inhibit specifically receptor activity. Such inhibitors may represent a novel therapeutic strategy for those cancers due to unregulated GFR activity.

References

Akrigg, D., Bleasby, A. J., Dix, N. I. M., Findlay, J. B. C., North, A. C. T., Parry Smith, D., Wootton, J. C., Blundell, T. L., Gardner, S. P., Hayes, F., Islam, S. A., Sternberg, M. J. E., Thornton, J. M., and Tickle, I. J. (1988). *Nature*, 335:745–746.

Baker, E. N. and Hubbard, R. E. (1984). *Prog. Biophys. Mol. Biol.*, 44:97–179.

Bargmann, C. I., Hung, M. C., and Weinberg, R. A. (1986). *Cell*, 45:649–657.

Bargmann, C. I. and Weinberg, R. A. (1988). *Proc. Nat. Acad. Sci. USA*, 85:5394–5398.

Bormann, B.-J., Knowles, W. J., and Marchesi, V. T. (1989). *J. Biol. Chem.*, 264:4033–4037.

Chothia, C., Levitt, M., and Richardson, D. (1981). *J. Mol. Biol.*, 145:215–250.

Coussens, L., van Beveren, C., Smith, D., Chen, E., Mitchell, R. L., Isacke, C. M., Verma, I. M., and Ullrich, A. (1986). *Nature*, 320:277–280.

Engleman, D. M. and Steitz, T. A. (1981). *Cell*, 23:411–422.

Gullick, W. J. (1988). In Cooke, B. A., King, R. J. B., and van der Molen, H. J., editors, *Hormones and their Action, Part II*, pages 349–360. Elsevier, Amsterdam.

Hanks, S. K., Quinn, A. M., and Hunter, T. (1988). *Science*, 241:42–52.

Livneh, E., Glazer, L., Segal, D., Schlessinger, J., and Shilo, B.-Z. (1985). *Cell*, 40:599–607.

Rothbard, J. B. and Taylor, W. R. (1988). *EMBO J.*, 7:93–100.

Sainsbury, J. R. C., Farndon, J. R., Needham, G. K., Malcolm, A. J., and Harris, A. L. (1987). *Lancet*, i:1398–1402.

Slamon, D. J., Clark, G. M., Wong, S. G., Levin, W. J., Ullrich, A., and McGuire, W. L. (1987). *Science*, 235:177–182.

Sternberg, M. J. E. and Gullick, W. J. (1989). *Nature*, 339:587.

Sternberg, M. J. E. and Gullick, W. J. (1990). *Prot. Eng.*, 3:245–248.

von Heijne, G. and Gavel, Y. (1988). *Eur. J. Biochem.*, 174:671–678.

Weiner, D. B., Liu, J., Cohen, J. A., Williams, D. V., and Greene, M. I. (1989). *Nature*, 339:230–231.

Yarden, Y. and Schlesinger, J. (1987). *Biochemistry*, 26:1443–1451.

Discussion

Q: How did you align the transmembrane sections?

A: There are various ways of aligning the transmembrane. One can either align them on the carboxyl end or the amino end or you can align them on the motif. This is not a family of closely homologous sequences. The extracellular domains are different in character and there are a different number of residues between the end of the transmembrane region and the tyrosine kinase domain. Therefore there could be different rotations for the different transmembrane region in relation to the extra- and intra-cellular domains. Thus the alignment is not of the same type as for the alignment of say active site residues in a family of enzymes.

Q: How were the statistics worked out for the frequencies of the residues?

A: The frequencies were those within the 20 transmembrane regions which are of course quite different from that in proteins in general.

Q: Is there experimental evidence to support your hypothesis?

A: After we proposed this specific model for dimerisation, there was a paper that showed that the Glu mutation in onc-neu promoted dimerisation (see Weiner *et al.* (1989) in reference list above). In general in the area of TK-GFR there is considerable body of experimental information that suggests that dimerisation is the mechanism of TK activation.

Q: Could there be a simpler explanation due to the presence of a charge in a hydrophobic environment?

A: The Glu mutation must be at residue 664. When it was introduced by site-directed mutagenesis at 663 or 665 the protein was not transforming. Similarly, a positive charge such as Lys or Arg at 664 also did not lead to transformation. One is therefore dealing with a specific stereochemical effect rather than simply a charge in the transmembrane region.

Q: Are you aware that the transmembrane region from glycophorin A can specifically inhibit the dimerisation of this protein?

A: After we proposed this model, we saw this work reported (Bormann *et al.*, 1989). This suggests that the transmembrane dimerisation may be a more general phenomenom than in just TK-GFR and the approach may have several applications.